Design and Production in Medieval and Early Modern Europe

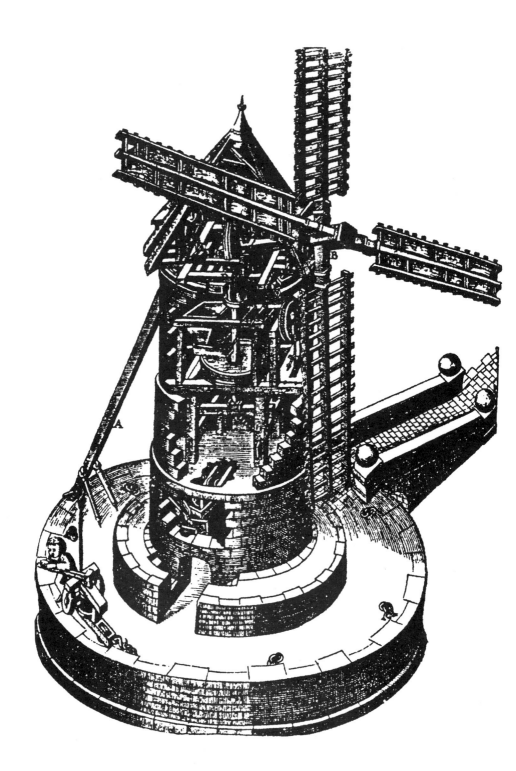

Wissenschaftliche Abhandlungen
Band LXII/4

Musicological Studies
Vol. LXII/4

Claremont Cultural Studies

Design and Production in Medieval and Early Modern Europe: Essays in Honor of Bradford Blaine

general editor

Nancy van Deusen

The Institute of Mediaeval Music
Ottawa, Canada

The Institute of Mediaeval Music
1270 Lampman Crescent
Ottawa, Canada
K2C 1P8

ISBN 1-896926-13-4

CONTENTS

ILLUSTRATIONS

Design and Production
in Medieval and Early Modern Europe:
Essays in Honor of Bradford Blaine

Mental designs are powerful. As Edward Grant points out in his contribution to this volume, what is *secundum imaginationem* achieves a result. This volume brings together, often in unexpected ways, the mental design that served as a catalyst, with the product in the mental and technical culture of what is commonly known as the Middle Ages. But, as the cultural historian of medieval technology, Lynn White, Jr., was often to point out, both designs and constructions that originated in or at least can be traced to the medieval period, were not by any means contained within that period. All too often, medieval "designs" remain influential today.

There is both precedent and powerful legitimation for what can be termed "technology and production" in the Middle Ages. For one, craftsmanship is given a prominent place within the early chapters of the Old Testament, in which workmanship is related, with success attributed to the influence of the Holy Spirit.

> And I have filled him with the Spirit of God, in wisdom, and in understanding, and in knowledge, and in all manner of workmanship, to devise cunning works, to work in gold, and in silver, and in brass, and in cutting of stones, to set them, and in carving of timber, to work in the manner of workmanship. (Exodus 31:3f)

This is one of the first—and few—times that the Spirit of God is explicitly mentioned in the opening chapters of the Pentateuch, and this certainly did not escape Augustine's notice. The final chapters of *The City of God* praise mental gifts and "an acuteness of intelligence" that results directly in both design and construction; the ability to form within the imagination, and the capacity to bring that conceptualization to a productive conclusion. Augustine writes:

> Just think of the progress and perfection which human skill has reached in the astonishing achievements of cloth making, architecture, agriculture, and navigation. Or think of the originality and range of what has been done by experts in ceramics, by sculptors, and by painters: of the dramas and theatrical spectacles so stupendous that those who have not seen them simply refuse to believe the accounts of

those who have. Think of the contrivances and traps which have been
devised for the capturing, killing, or training of wild animals; or again,
of the number of drugs and appliances that medical science has
discovered in its zeal for the preservation and restoration of men's
health; or again, of the poisons, weapons, and equipment used in wars,
devised by military art for defense against enemy attack; or even of the
endless variety of condiments and sauces which culinary art has
discovered to minister to the pleasures of the palate. It was human
ingenuity, too, that devised the multitude of signs we use to express and
communicate our thoughts—and, especially speech and writing. The
arts of rhetoric and poetry have brought delight to men's spirits by their
ornaments of style and varieties of verse; musicians have solaced
human ears by their instruments and songs; both theoretical and applied
mathematics have made great progress; astronomy has been most
ingenious in tracing the movements, and in distinguishing the
magnitudes, of the stars. In general, the completeness of scientific
knowledge is beyond all words and becomes all the more astonishing
when one pursues any single aspect of this immense corpus of
information.[1]

All of the categories of "design" Augustine has brought to our attention are
included in this volume. In his contribution, "The Nature of Western European
Science in Late Middle Ages," Edward Grant sets forth hypothetical, imaginative
constructs, as he states (page 8), "It would be no exaggeration to claim that the
medieval Aristotelian-Ptolemaic cosmos was largely a series of mental constructs
fashioned from metaphysics and theology. Among the most important, and even
foundational mental constructs I would include: prime matter, angels and
intelligences, celestial orbs . . . , God's infinite immensity and omnipresence,
impressed forces, prime mover, celestial influence, universal nature, celestial
corruptibility, hierarchical nobility of celestial bodies, and virtual qualities." The
panorama of mental constructs is comprehensive. Nearly all imaginable areas of
interest are accessed by one or another of these questions. Further, Grant brings
up the problem of mental constructs and "reality," i.e. what can be observed or
perceived with the senses.

In his article, "Agricultural Practice and Garden Design in Renaissance
Liguria," George Gorse explores basic imaginary constructs such as "the
sublime," (and some of the first articulations of the sublime) as well as "the
Romantic binary construction" of categories such as "feudal" and "modern," thus
posing the question of what mental constructs we ourselves have inherited, and

---

[1] Augustine, *The City of God*, trans. Gerald G. Walsh, S. J., Demetrius B. Zema, S.J., Grace
Monahan, O. S. U., Daniel J. Honan (New York, 1950, repr. 1958), pp. 526f.

how, for example, binary constructions produce perception. He states, "Representation *is* functional and it is real in creating (and recreating) social orders" (p. 40).

Richard W. Unger, in "Technology Illustrated: Problems of Sources," explores technological products and their efficiency, as well as the relationship of these products to visual representation—in itself, certainly, a product, as he states. What do illustrations of instruments of all kinds reveal? How reliable are they as delineatory for the tasks to be performed by, or, perhaps, upon those instruments? What are these illustrative, delineatory, "products" actually "good for" are some of the questions Unger brings up in his intriguing essay. The problems with sources he expresses are the very issues historians of technology have in common with musicologists focusing on pictures and diagrams of music instruments in order to determine how music was performed upon these instruments, the status of musicians performing—or qualities of the music of the period itself. A complex medium of representation is called into discussion for instruments of *any* kind, but the potential for this representation is great.

Elspeth Whitney takes on the problematic intersection of the mechanical arts and moral philosophy in her essay, "*Artes Mechanicae*, Craftsmanship and Moral Value," a topic to which Bert Hall and Ranald Mackenzie Macleod return in the concluding essay of this volume. Joseph Donatelli explores interrelationships between "design," "production," namely the close connection and "deeply ambivalent relationship between writing and printing during the fifteenth and sixteenth centuries," demonstrating, as he states, both the complementary nature and accommodation of media during this period.

In a trio of papers concerned directly with the concept of *figura*—a pivotal concept that lurks in all of the articles—Richard Wingell sets up the medieval equivalence of figural shape as "mental and physical construct of heard sound." Nancy van Deusen shows how the mental design of a "stringed instrument" continued, throughout the medieval period, to produce structure within medieval exegesis, pointing to a revolutionary break with this structure and expectation on the part of its reception. Carol D. Lanham, again, demonstrates the power of figural, essentially conceptual, images in communication.

Finally, Bert Hall and Ranald Mackenzie Macleod's article, "Technology, Ecology, and Religion: Thoughts on the Views of Lynn White," explores, first of all, the underlying, at its inception mental, construct of the academic essay within the production of the highly-influential historian of medieval technology, Lynn White, Jr., who was also Brad Blaine's mentor. This final contribution also, however, clearly presents a feature Lynn White and Brad Blaine share, namely, a conviction that mental constructs hold undeniable consequences for social action—and bear responsibility for an outcome within a real world.

Thus, this volume, in honor of Brad Blaine, reflects not only Professor Blaine's own field of intellectual enquiry, namely, the scientific imagination and medieval technology, but his own very clear mental construct of magnanimous collegiality that has resulted in many and diverse "products" throughout his academic career. It is also Brad Blaine's clear vision of teaching and learning that has had an immediate as well as long-term effect in terms of the students who crowd his classes and who also continue their studies at graduate universities—as well as, in fact, the rest of their lives.

Many of these essays were presented during a conference of the same title, held in Professor Blaine's honor on the Scripps College campus, October 1993. Grateful acknowledgment for the support of this conference and Festschrift is given to Scripps College, as well as the Claremont Consortium for Medieval and Early Modern Studies. Appreciation is also expressed to Nancy Bowen, Ph. D. candidate in Medieval Studies for her help in the production of this volume.

NANCY VAN DEUSEN
Claremont Graduate University
Fall, 1997

CONTRIBUTORS

NANCY VAN DEUSEN
    Claremont Graduate University

JOSEPH M. P. DONATELLI
    University of Manitoba

GEORGE L. GORSE
    Pomona and Scripps Colleges

EDWARD GRANT
    Indiana University

BERT HALL
    University of Toronto

CAROL D. LANHAM
    UCLA Center for Medieval and Renaissance Studies

RANALD MACKENZIE MACLEOD
    University of Toronto

JAMES OTTE
    University of San Diego

RICHARD W. UNGER
    University of British Columbia

ELSPETH WHITNEY
    University of Nevada, Las Vegas

RICHARD J. WINGELL
    University of Southern California

The Nature of Western European Science in the Late Middle Ages

by Edward Grant

MEDIEVAL SCIENCE: THE BEGINNINGS

By the twelfth century, western Europe had developed a hunger for new secular learning. Up to that time, what scholars knew about the physical world was derived from traditional Latin handbooks that contained the remnants of a popular science that went back to the Greeks of the Hellenistic period. The knowledge they sought has been appropriately characterized as "Greco-Arabic" science because it consisted of works written in Greek within a Greek cultural orbit going back as far as the fifth century B.C., and also of works written in Arabic that had been either translated from Greek or were original compositions. The number of works translated from Arabic into Latin far exceeded those translated from Greek into Latin. These translations were made by scholars from all parts of Europe, who went to Spain, Sicily, and northern Italy, or were already inhabitants of these places. Most of those who translated from Arabic to Latin had to learn Arabic, which was not their native language. Without their extraordinary achievements, late medieval science in Europe might never have occurred.

This vast amount of new learning that entered western Europe, and which had never before been known in the Latin language, is appropriately divisible into two categories: the first includes treatises that were devoted to technical science, such as Euclid's *Elements* and Ptolemy's *Almagest*; the second embraces those that were classifiable as works of natural philosophy, especially those written by Aristotle (along with commentaries on Aristotle's treatises by the Arabian commentators, Averroes [1126–1198] and Avicenna [980–1037]).

Although both of these divisions of Greco-Arabic science were important for the development of the history of science, I will argue that what medieval scholars did with natural philosophy and the role they assigned to it in intellectual life was ultimately more important than what they did with the technical sciences. In this paper, I shall focus on the role of natural philosophy.

WHAT WAS NATURAL PHILOSOPHY DURING THE MIDDLE AGES?

The subject matter of natural philosophy *(philosophia naturalis)*, or natural science *(scientia naturalis)*, as it was sometimes called, was a study of nature, or the world, which Aristotle had divided into two radically distinct regions, the terrestrial and celestial. In its broadest sense, natural philosophy was the study of bodies undergoing change and motion. It was meant to include the study of the four kinds of change that Aristotle had distinguished in the world, namely changes in substance, quality, quantity, and change of place, or as we would call it, motion. But it was even broader than that because it also included the study of an incorruptible substance, the celestial ether, from which the planets, fixed stars, and the orbs that carried them were made. But the celestial ether also suffered some change, because it moved naturally around the heavens with a circular movement. Aristotle had studied the operation and structure of the physical world, or nature, in a series of works that were known as the "natural books," which primarily included those treatises bearing the titles *Metaphysics, Physics, On the Heavens, On Generation and Corruption, Meteorology*, and *On the Soul*. Natural philosophy was mostly a study of bodies undergoing change and motion, and it remained so during the Middle Ages and into the eighteenth century, when, in 1726, Isaac Newton retained the expression "natural philosophy" in the title of the third edition of his monumental work, *Mathematical Principles of Natural Philosophy*.

During the late Middle Ages in Europe, natural philosophy, or the study of nature, was essentially a study of the relevant works of Aristotle on which medieval scholars wrote commentaries. It was therefore an essentially "bookish" enterprise and, because of the way natural philosophy was taught and the place where it was taught, it soon assumed the appearance of a traditional and almost permanent activity. This happened because natural philosophy—and by that I shall mean a natural philosophy based on Aristotle's natural books—became the core curriculum of the universities of the Middle Ages. In fact, the medieval universities emerged in part, perhaps for the most part, because the works of Aristotle presented them with a large, cohesive body of knowledge about nature that enabled them to develop a suitable curriculum for the degree of master of arts. Without Aristotle's works on natural philosophy and logic as the centerpiece of medieval higher education, it is difficult to imagine the sort of history that universities would have had.

The first of the universities of medieval Europe—the universities of Bologna, Paris, and Oxford—were already in existence by around 1200. By this time, western Europe had come into possession of much, if not most, of the cumulative scientific knowledge that had been produced by the civilizations of Greece and Islam. When compared to the meager number of scientific texts available just prior to the age of translation, the achievements of the translators are truly staggering. The translations enabled two things to occur that might not otherwise have happened. First, they made available a powerful and comprehensive body of scientific literature which provided the curriculum on which the medieval university was dependent. And, second, that curriculum, with its core of natural philosophy in which was contained a mass of ideas and theories about the cosmos, provided a foundation for the continuous development of science from the thirteenth century to the present.

By 1200, when the first universities had just emerged from the shadows of history, the works of Aristotle, along with Averroes' commentaries on those works, were already regarded as the most plausible basis for a university curriculum. But natural philosophy is a subject that is always potentially problematic for established religions, especially those that have a sacred scripture. And Aristotelian natural philosophy was especially problematic for

Christianity because in the form Aristotle gave it, natural philosophy provided a detailed system of explanation, complete with metaphysical principles, that served, even if innocently, as a rival world view to what was expressed in the Bible.

As if that were not enough, there were specific points of friction and controversy. Among these the most dangerous for any religion with a creator God was Aristotle's powerful arguments in favor of the world's eternity and therefore against the concept of a temporal creation. But there were other difficulties with Aristotle's natural philosophy. His views on the immortality of the soul, which he seemed to deny, were unacceptable, as were his ideas about the impossibility of creation from nothing, which formed part of his battery of arguments favoring the eternity of the world. Belief in the resurrection of the body was also unacceptable in Aristotle's philosophy. Theologians also found his conception of God objectionable. In Aristotle's depiction of the cosmos, God was an inactive abstraction who was totally divorced from the world, who neither created it nor knew anything about it and consequently had no concern for it.

We are not surprised to learn therefore that some opposition developed almost immediately to the works of Aristotle. Efforts were made to ban his works at the University of Paris in 1210 and 1215, and in 1231 there was even an attempt to purge them of offensive ideas. All such efforts were in vain, and by 1255, Aristotle's works had become the official core of the arts curriculum at Paris. There were further troubles in the 1260s and 1270s when conservative theologians, inspired by Saint Bonaventure, were alarmed at the great emphasis placed on Aristotle's secular philosophy. They were also fearful at the possible damage that might result if Aristotle's philosophical ideas and methodology were applied to Christian theology. Because repeated warnings of the inherent dangers of secular philosophy and the perils of its application to theology had been ignored, traditional, conservative theologians convinced the bishop of Paris, Etienne Tempier, to take action. In 1270 he condemned 13 propositions drawn largely from the ideas of Aristotle and Averroes and in 1277 proclaimed a massive condemnation of 219 proposi-tions, the acceptance of any one of which would result in excommunication.

Despite the troubling aspects of Aristotle's natural philosophy and the adverse theological reaction to some of his ideas at the University of Paris during the thirteenth century, his works were welcomed by almost all natural philosophers and theologians in the Latin West. Aristotle had too much to offer. He was the principal guide in understanding the operations of the cosmos. Caught between fears about possible dangers to the faith on the one hand, and the feeling that Aristotle's metaphysics and physics were instrumental for understanding nature on the other hand, medieval Christians chose to embrace Aristotle with the hope that the potential dangers to their faith would never materialize. It was a momentous decision. Although the Christian faith eventually would be diminished by the triumphs of natural philosophy and science, a consequence that no one in the Middle Ages could have foreseen, it is nevertheless a remarkable feature of Western intellectual history that theologians respected and embraced Aristotle's works. Their approval of Aristotle's natural philosophy was undoubtedly a powerful contributing factor to the success of natural philosophy in the West. Perhaps one might even wish to argue that the cooperation of the theologians was indispensable.

THE MEDIEVAL UNIVERSITY

Without the medieval university, however, natural philosophy might never have taken root in Western thought. The enormous success of natural philosophy in the Middle Ages was virtually guaranteed when the universities of Europe adopted the "natural books" of Aristotle as the basis of the curriculum for the master of arts degree. For students who wished to teach at the university level, it was necessary to obtain a teaching license *(licentia docendi),* which was a formal prerequisite for becoming a master of arts and usually required an additional two years of course work. To achieve this goal the student was required to attend lectures on natural philosophy and to take some courses in the technical sciences, or quadrivial arts, as they were known, which embraced astronomy, geometry, arithmetic, and music. The ultimate goal of a full arts education was to convey an in-depth knowledge about the structure and operation of the physical cosmos. Students at medieval universities were essentially exposed to a physics-cosmology-logic curricu-lum. That curriculum served as the regular fare for legions of students who passed through the universities of Europe between the thirteenth and the seventeenth centuries. From this, we should rightly infer that the popular depiction of the Middle Ages as a period largely devoid of science and openly hostile to it is a caricature and distortion of the truth. The reality is quite otherwise. Interest in the structure and operation of the physical world was intense and widespread, and manifested itself at the universities in the study of Aristotelian natural philosophy. Indeed, despite the advent of the Scientific Revolution, the reformed universities that emerged as a result of its ultimate influence, and that now function in the modern world, have never again seen fit to make science the culmination of an arts education.

But how was natural philosophy taught in the medieval universities? Two basic methods were employed to teach natural philosophy, and each method had its counterpart in the scholarly literature. The less widely-used method was the commentary. Since lectures were at first largely sequential, section-by-section expositions of an authoritative text (for example, this or that work of Aristotle), the analogous written commentary simply followed the same procedure. A section of text was read and then explained with occasional elaborations by the commentator; upon completion, the next section or unit of text was read and expounded; and so on. It was not unusual in these commentaries to provide references to other texts in support of an argument.

But the most widely and regularly used format for medieval cosmology, and for natural philosophy in general, was the *questio,* or question format. The concept of a medieval "scholastic method" is more intimately associated with the *questio* form of literature and analysis than with anything else. The practice of dealing with questions may have arisen from the commentary. When a master had finished reading and explaining a number of passages or sections of an

Aristotelian work, it became customary at the end of a lecture to pose a question based on those passages and to present the pros and cons of the answers to it, followed by a solution. In time, the questions previously posed at the end of a lecture came to displace the commentary on the text itself. Eventually the teaching masters focused on specific questions or problems that followed the order of the required text and developed from it. Collectively, these questions formed the totality of the master's formal lectures on that particular Aristotelian work. The written forms of this pedagogical technique that have survived— questions treatises, or *questiones*—are often associated with the names of well-known masters, who presumably presented as lectures some version of the surviving text.

During the Middle Ages and even into the seventeenth century, scholastic natural philosophers centered their teaching and analysis of authoritative texts around the *questio*. Natural philosophy, which constituted the core of the curriculum, was thus taught by the analysis of a series of questions posed by a master and eventually determined—that is, resolved—by him. Collections of these *questiones* on the different works of Aristotle, and other works as well, were written down by the masters and later copied at the university bookstore where copies could be rented or sold to students and masters.

Although occasional variations were made in the arrangement of the constituent parts of the *questio*, scholastics tended to present their arguments in a rather standard format that remained remarkably constant over the centuries. First came the enunciation of the problem or question, usually beginning with a phrase such as "Let us inquire whether" or simply "Whether" *(utrum)*: for example, "Whether the earth is spherical," or "Whether the earth moves," or "Whether it is possible that several worlds exist." This was followed by one or more— sometimes five or six—solutions supporting either the negative or affirmative position. If arguments for the affirmative side appeared first, the reader could confidently assume that the author would ultimately adopt the negative position; conversely, if the negative side appeared first, it could be assumed that the author would subsequently adopt and defend the affirmative side. The initial opinions, which would ultimately be rejected, were called the "principal arguments" *(rationes principales)*.

Immediately following the description of the principal arguments, the author might further clarify and qualify his understanding of the question or explain particular terms in it. With the necessary qualifications completed, the author was ready to present his own opinions, usually by way of one or more detailed conclusions or propositions. In order to anticipate objections, the master might choose to raise doubts about his own conclusions and subsequently resolve them. To conclude the question, he would respond, in sequence, to each of the "principal arguments" enunciated at the outset.

By its very nature, the *questio* form encouraged differences of opinion. It was a good vehicle for dispute and argumentation. Medieval scholastics were trained to dispute and therefore often disagreed among themselves. In almost all questions, there were at least two, and often more, previously distinguished positions. Had the *questiones* been of a form in which a single conclusion was provided which was then defended and upheld by a series of arguments, medieval and Renaissance scholasticism would have been less interesting and productive than it was.

The scholars who wrote the *questiones* were masters of arts, or else masters or doctors of theology. They were trained to evaluate arguments critically and by a process of elimination arrive at the most plausible response under the circumstances. Their skill and ingenuity were displayed by the introduction of subtle distinctions that, upon further development, might yield new opinions on this or that question.

Hundreds of questions were formulated from Aristotle's natural books and collectively formed the basis of medieval natural philosophy. To a considerable extent, doing medieval science meant analyzing and evaluating these questions and arriving at satisfactory conclusions and answers. What kinds of questions did medieval natural philosophers consider? The questions ranged over every aspect of the world's operations and activities. They asked whether there were other worlds besides ours and whether an infinite space might lie beyond our world; and they also inquired whether our world was eternal or created and whether it was perfect. With regard to the celestial region, they wanted to know how many spheres existed and what was the order of the planets. They also inquired whether contrary qualities such as light and heavy, or dense and rare, actually existed in the heavens and they asked what caused the celestial orbs to move in never-ending circular motion and whether the planets receive their light from the Sun or are self-luminous. They were concerned about the influence of the celestial region on the terrestrial, and posed numerous questions about the nature of the terrestrial region, which seemed so strikingly different from the celestial realm. For example, they asked whether elements remain or persist in a compound? Are there only four elements? Is there any pure element? Can one element be generated directly from another? They also inquired about the earth, asking, for example, whether it was spherical, or at rest, or wholly habitable. Problems of motion were frequently proposed: for example, whether natural motion is quicker at the end of a motion than at the beginning and whether projectile motion is quicker in the middle of its trajectory than in the beginning or end.

Over the centuries, scholastics departed from Aristotle on many issues. What is curious, however, is that these departures were rarely acknowledged. The prevailing tendency was to reconcile one's conclusions with Aristotle's interpretation, which was itself often in dispute. Departures were frequently made by way of an interpretation that seems to us quite different from anything Aristotle could have had in mind, and yet, so great was the authority of Aristotle that scholastic authors usually avoided the appearance of disagreement, even when it seems apparent to us that they were expressing a quite different interpretation than Aristotle. Thus we are left to ponder whether they were consciously aware of their departure but opted to remain silent out of respect for Aristotle's authority, or whether they believed that their interpretation was Aristotle's genuine opinion. Thus Aristotle not only furnished the sequence of questions, by virtue of the order in which he chose to consider the subjects in his works, but his powerful authority made it difficult to proclaim new departures.

Although the question form was the most important in scholastic literature, there were other kinds of treatises in which some theme in natural philosophy was presented in a systematic way, progressing from beginning to end. Such treatises usually included the word *tractatus*, that is, treatise, in the title. Nicole Oresme (c. 1320–1382) was perhaps the most unusual in this regard, writing a number of important tractates (for example, *Treatise on the Configuration of Qualities and Motion*, *Treatise on the Commensurability or Incommen-surability of the Celestial Motions*, and *Treatise on Ratios of Ratios*), in addition to at least five questions and commentaries on the works of Aristotle.

An interesting feature of medieval Latin natural philosophy was frequent use of hypothetical arguments, which were sometimes described by the Latin expression *secundum imaginationem*. Such arguments were often derived from medieval conceptions of God's absolute power and were associated with certain articles condemned in 1277. When the bishop of Paris condemned these articles he did so in order to emphasize that some of Aristotle's arguments about the world were not to be interpreted as unqualifiedly true. Thus Aristotle had argued that it was impossible that more than one world could exist. But article 34 of the Condemnation of 1277 condemned the idea that God could not create more than one world. Hence it was necessary to concede that more than one world was possible because God had the power to make other worlds, if He wished. As a consequence, some scholastic authors argued that other worlds might exist because God could have made them, even though they were convinced that God had created only our world, which they believed was unique. In effect, they were compelled to concede that God could do something that Aristotle had thought was impossible. The same situation arose in connection with the possibility of the existence of a vacuum or vacua. Aristotle had denied even the possibility of the existence of a vacuum anywhere. But the Condemnation of 1277 made it necessary to concede that God could create a vacuum, or many vacua, wherever He pleased. In such arguments God was frequently imagined to produce a vacuum by annihilating all or some of the matter within our world. Many questions, in the form of "thought experiments," were then posed about the behavior of bodies in such empty spaces.

The formulation of counterfactuals formed an integral part of medieval science. Although the clash of natural philosophy and theology was responsible for some of it, many hypothetical theses *secundum imaginationem* were developed within the context of natural philosophy. The intension and remission of forms was perhaps the most significant. It was an effort to quantify what seemed unquantifiable, that is, to quantify the intensities of variable qualities, such as color, temperature, sound, health, pain, accidents of the soul such as cognition and imagination, passions of the soul such as fear, hope, sorrow and joy, and so on. In the process, not only were graphing techniques developed, but certain important kinematic definitions and theorems were proposed and demonstrated, namely the definitions of uniform and accelerated motion, which were then used to demonstrate the famous mean speed theorem and which Galileo used as the foundation of his new mechanics. But as is also well known, the fourteenth-century authors of these important concepts did not apply them to the motion of real bodies. They utilized these important ideas in imaginary exercises. When scholastic authors applied mathematics and logic to real and imaginary problems, they did so to test their ability to maintain logical consistency, not to arrive at, or test, physical truths. It required a Galileo to apply the mean speed theorem to real falling bodies and to devise the inclined plane experiment to determine its validity.

If hypothetical conditions played a significant role in medieval natural philosophy, we might also inquire about the role of mental constructs. It would be no exaggeration to claim that the medieval Aristotelian-Ptolemaic cosmos was largely a series of mental constructs fashioned from metaphysics and theology. Among the important, and even foundational, mental constructs I would include: prime matter, angels and intelligences, celestial orbs (concentric, eccentric, and epicyclic), God's infinite immensity and omni-presence, impressed forces, prime mover, celestial influence, universal nature, celestial incorruptibility, hierarchical nobility of celestial bodies, and virtual qualities. By contrast, the four elements would not qualify as mental constructs because they appear to have been derived from gross observation. Most medieval concepts, however, seem to be classifiable as mental constructs, because their *raison d'être* lies overwhelmingly in metaphysics or theology, rather than in experience or observation.

In applying the notion of "mental constructs" to medieval concepts, however, we must proceed with extreme caution. Hindsight enables us to determine that the concepts just mentioned had no basis in experience and were in fact mental constructs. But they were not regarded as mental constructs in the Middle Ages. To medieval natural philosophers, they were as real—nay, in many instances more real—than the silver and gold coins they received as remuneration for their teaching efforts or the parchments on which their degrees were written. We can readily see this in the medieval attitude toward intelligences, or angels as movers of celestial orbs. Aristotle had already invoked spiritual intelligences as real movers of celestial orbs. Angels were a reality from the early days of Christianity. It took Avicenna and Moses Maimonides to conflate Aristotelian intelligences with Koranic and biblical angels, respectively. Although it was not unanimous, the Latin West adopted the identification of angels and intelligences and assumed that these spiritual substances were external motive causes for the motions of the celestial orbs. Most scholastic natural philosophers probably believed that an angel or intelligence was more real—because more noble and perfect—than any terrestrial object. In the Middle Ages, it was easy, and even natural, to believe that invisible spiritual substances were the movers of invisible, trans-parent celestial orbs. The medieval physical world depended for its operations on invisible substances and powers that were invoked and developed as needed. Indeed, even the ultimate basis of material objects, prime matter, was itself an entity whose existence had to be assumed so that the four elements, and the visible compounds derived from them, could be explained.

THEOLOGIAN-NATURAL PHILOSOPHERS

In the development and evolution of natural philosophy during the late Middle Ages, theologians played a major role. Indeed many of them might be appropriately described as "theologian-natural philosophers." Because the theologians welcomed natural philosophy and studied it in the course of their careers, and some even wrote treatises on the subject, science and natural philosophy became highly respected disciplines. It is important to realize that virtually all who sought the degree of master or doctor of theology had already acquired a master of arts degree, or its equivalent, and were therefore familiar with Aristotle's natural philosophy.

The unusual development that produced a whole class of theologian-natural philosophers serves as a key for understanding the fate of science and natural philosophy in medieval western Europe. The absence of significant strife between theology and science is, in my judgment, attributable to the emergence of a class of theologian-natural philosophers. Because they were trained in both natural philosophy and theology, medieval theologians were able to interrelate science and theology with relative ease and confidence. In this important activity, they rarely exposed themselves to possible excommun-ication. Occasional reactions against natural philosophy, as in the early thirteenth century when Aristotle's works were banned for some years at Paris, and in the later thirteenth century when the bishop of Paris issued the Condemnation of 1277, ought to be interpreted as relatively minor aberra-tions when viewed against the grand sweep and scope of the history of Western Christianity.

Christianity provided a sympathetic environment for the sustenance and development of natural philosophy and science. It posed few obstacles to the practice and development of science. In fact, by allowing natural philosophy to form the graduate curriculum in the medieval universities, medieval Christianity showed that it was prepared to do more than merely tolerate natural philosophy's existence. It actively promoted natural philosophy in an open and public way.

But even if Western Christianity had only tolerated the study and use of natural philosophy, that would have been a sufficiently significant contribution. For we can easily imagine a scenario in which hostile religious authorities might have posed serious obstacles to the pursuit of science and natural philosophy. In a society where religion was fundamental, the hostility of the theologians, or perhaps even their lack of positive interest, might have proved fatal. During the Middle Ages, the Catholic Church avoided this course of action. Had they done otherwise, the creative scientific and philosophic activity that occurred within Catholic countries during the sixteenth and the seventeenth centuries would have been impossible. If the Church's attitude had been hostile to natural philosophy, and to science in general, they would have moved against science on a broad front instead of restricting their displeasure to Copernicus's heliocentric astronomy.

But if the theologian-natural philosophers were instrumental in preventing a potentially destructive clash between theology and natural philosophy, the medieval university provided the essential institutional base that enabled natural philosophy to play a prominent role in European intellectual life. Indeed, because natural philosophy and theology were both taught in many universities, it was the place where both disciplines co-existed, and the place where natural philosophy was gradually transformed into experimental and mathematical science between the seventeenth and the nineteenth centuries.

Derived almost wholly from the Greco-Arabic legacy that entered western Europe for the first time in the late twelfth and the thirteenth centuries, medieval natural philosophy was itself a completely new intellectual phenomenon which had never before been seen in western Europe. It should be viewed as the true beginning of modern science in western Europe, not because of any significant achievements that emerged from it in the Middle Ages—although there were certainly some interesting developments—but because medieval natural philosophy established vital habits of scientific inquiry and also managed to formulate most of the major problems that would be solved in the seventeenth century.

Centuries of analysis, evaluation and criticism had altered the Greco-Arabic inheritance in natural philosophy and produced significant additions and alterations. While Aristotle remained the supreme authority, his views and judgments were challenged on a number of issues, especially with respect to problems of motion. These traditional challenges to Aristotle ought to be viewed as part of the legacy that was bequeathed to those non-scholastics who produced the new science. That most of these medieval ideas were eventually rejected should not minimize the important and vital role played by the range of knowledge that we categorize as medieval natural philosophy. It was a highly sophisticated body of learning about the physical world which generated discussion and controversy for more than four centuries. Because it formed the curriculum of the university, which was firmly and permanently rooted in medieval society and in subsequent Western history, a large body of natural philosophy was taught to thousands of students generation after generation. Not only did natural philosophy establish a basic world view, but it produced and disseminated an approach to nature that emphasized certain habits that were beneficial to Western science. The *questio* form encouraged differences of opinion by its very nature, and was a good vehicle for dispute and argumentation. Scholastics in the Middle Ages often disagreed among themselves because of their training in dispute and argumentation. Far from slavish devotion to Aristotle, they were emboldened by the very system within which they were nurtured to arrive at their own conclusions, or at the very least to choose from a minimum of two, and often more, previously distinguished conclusions. The means for deciding between one conclusion or another was based on reason and/or metaphysics. What was largely absent from medieval natural philosophy were helpful ways to choose between alternatives, namely critical observation and a systematic and regularly employed experimental method.

Despite these glaring deficiencies, which the ancient Greeks shared, natural philosophy during the late Middle Ages was a positive enterprise which shaped Western attitudes toward the physical world by posing a very large number of questions to nature. Indeed, posing questions to nature was the most essential feature of medieval Aristotelian natural philosophy. When the answers of Aristotelian natural philosophers to the more important questions about that world were found wanting in the sixteenth and seventeenth centuries, different and more satisfying responses were proposed that changed our understanding of the world and altered the nature of science itself. These dramatic changes were made possible by the natural philosophy developed in the late Middle Ages. The way for a medieval master to do science within the framework of the university system was to present an oral or written analysis of some of the many questions that collectively formed the basis of the medieval view of the physical world. The master was expected to critically evaluate each question and arrive at a conclusion, usually one that had traditional support, but which occasionally represented a new departure. In principle, natural philosophers were expected to evaluate arguments critically and, by a process of elimination, arrive at the truth. Centuries of disputation produced a variety of opinions on a very large number of questions. Some of these questions and answers played a pivotal role in the emergence of the new science.

But those specific questions and answers are not the basis for the medieval contribution to the emergence of a new non-scholastic science. The medieval practitioners of natural philosophy prepared the way for the new science of the seventeenth century by inculcating a spirit of inquiry and by establishing the concept of reasoned argumentation and acceptation of divergent opinions about the nature of the physical world, even when some of those opinions occasionally clashed with traditional theological interpretations. Long before the unfortunate Galileo affair, medieval natural philosophers had created an environment that was favorable to scientific inquiry, and had also identified a whole series of important questions that they had vainly tried to answer over the centuries. Although the answers to these problems, supplied by the likes of Copernicus, Galileo, Kepler, Descartes, and Newton, demanded the repudiation of medieval Aristotelian natural philosophy, the latter had already made its contribution to the advance of science. It had done its work, and a basically new natural philosophy, called "modern science," was brought into being to replace it.

THE TRADITIONAL SCIENCES

I should like to discuss the traditional, technical sciences that also played an important role in preparing the way for the advent of modern science. As I have already mentioned, the second aspect of the arts curriculum was the old quadrivium of the seven liberal arts, which consisted of arithmetic, geometry, astronomy, and music. The versions of these subjects that were taught in the medieval universities differed vastly from what passed for these subjects in the time of Boethius and the early Middle Ages. In addition to the science texts used in the arts curriculum, there were medical treatises used in the medical school. Most of the treatises taught in the universities were composed in the thirteenth and fourteenth centuries or were translations of Greek or Arabic treatises that were unknown in the earlier Middle Ages: Euclid's *Elements*, Sacrobosco's

Treatise on the Sphere, treatises on statics by Jordanus de Nemore, works on music by Johannes de Muris, and treatises on optics or perspective drawn from works by Ptolemy Alhazen, John Pecham, Roger Bacon, and others. Works by Galen, Averroes, Avicenna, and others con-stituted the fundamental texts used in the medical schools.

During the course of the Middle Ages, however, numerous scientific treatises were written on the whole range of the physical, medical and biological sciences, usually by university-trained scholars. Although relatively few of these treatises were used as university texts, they serve as representative achievements of medieval scientists and natural philosophers. Most of these were not *questiones* but straight-forward treatises that tended to follow the format generally used in that particular field. Thus medieval scientists reacted essentially to the textual tradition they inherited, and remained within it. Optical and astronomical treatises, for example, were overwhelmingly theoretical, and reveal little by way of experiments or critical observations. Doing science within this genre of scientific works was as internal and tradition-bound as were the *questiones* treatises on the works of Aristotle. Their empirical content was largely drawn from the texts they used and depended on. Few repeated experiments or observations, but relied on what they found in their authorities. If the authorities conflicted, the solution rarely involved an empirical test or experiment to decide the issue, even where it might be feasible to do so.

But there were exceptions. A few important treatises were written in the Middle Ages that reveal careful observation and even designed experiments. In this regard one need only mention the letter of Peter Peregrinus *On the Magnet* (c. 1269), the great treatise on falconry by the Holy Roman Emperor, Frederick II, the zoological observations of Albertus Magnus (in books 20–26 of his *Twenty-Six Books on Animals*), and perhaps the greatest work of all, Dietrich of Freiberg's treatise *On the Rainbow (De iride)*, where Dietrich was the first to show that in the production of the rainbow, light is reflected internally within each raindrop and that the colors of the secondary rainbow are the reverse of those in the primary bow. These works show keen observation and, with Dietrich at least, a good sense of careful experi-mentation. But such works were relatively rare in the Middle Ages. Designed experiments and systematic observation and testing were never regarded as integral features of scientific investigation. Although Aristotle's opinions and reports were frequently challenged during the Middle Ages, much of what he and other authorities said was usually considered the equivalent of personal observation and assumed to be true. Few were inclined to re-examine or reaffirm the claims of respected authorities.

A fascinating illustration of the medieval approach to experiment appears in the scholastic defense of the concept that "nature abhors a vacuum." In commentaries and *questiones* on Aristotle's *Physics* and *On the Heavens* and occasionally in other treatises, scholastics presented a considerable array of experimental evidence drawn from a variety of Greco-Arabic sources to prove "conclusively" that nature abhors a vacuum and would do virtually anything to prevent its formation. They spoke of a candle burning over water in an enclosed vessel; of the exhaustion of air from siphons and reeds; of the common bellows, the sides of which could not be separated by any force once the air had been exhausted and the sides had come into contact; of the clepsydra that would not permit its water to fall through the tiny holes on its bottom for fear that a vacuum would remain in the vessel; and of the impossibility of two perfectly plane surfaces to be separated if separation implied the creation of a vacuum. All of these experiences were capable of explanation by appeal either to a vacuum or a plenum. Because the actual existence of any vacuum, large or small, was wholly contrary to Aristotelian physics and cosmology, medieval explanations of these experiments always assumed that "nature abhorred a vacuum." As one means of denying the possibility of a vacuum, medieval natural philosophers invoked a continuous, material medium, and when that seemed threatened, they invoked a "universal nature" that was supposed to prevent disruption of that continuity and thereby prevent formation of the dreaded vacuum.

At least one of these experiences was clearly based on reason and imagination and was incapable of demonstration. The separation of plane surfaces was originally proposed by the Roman poet Lucretius *(De natura rerum)* as proof of the existence of vacua, for, as he put it, "if two bodies suddenly spring apart from contact on a broad surface, all the intervening space must be void until it is occupied by air."[1] During the Middle Ages, scholastic authors, who accepted Lucretius's argument, got around it by insisting that any two surfaces in perfect contact and without air or any other medium intervening could not possibly be separated. It followed then that surfaces that were separable could not have been in perfect contact.

But experiments involving bellows, clepsydras, and burning candles were experiences that might have been observed, even by scholastic authors. However, given the universally accepted Aristotelian assumption that "nature abhors a vacuum," direct observation of such experiments would not have altered this deep conviction. Explanations in defense of Aristotle's claim were readily available. Such explanations were thought to be of great importance because the existence of a vacuum was completely subversive of Aristotelian cosmology and physics.

With the possible exception of the bellows, none of these "experiments" and experiences were devised in the Latin Middle Ages, but were an inheritance from Greco-Arabic scientific treatises. Although frequently cited, there is no evidence that these experiments were actually repeated by those who cited them. They had no reason to challenge their validity, although they might embellish them, as when John Buridan declared that not even twenty horses, ten on each side, could pull apart the sides of a completely stopped up and evacuated bellows. It was Buridan's emphatic way of declaring that nature would not permit any force, however great, to create a vacuum. These important experiments would not be re-examined critically until the sixteenth and seventeenth centuries, when, ironically, they were invoked in order to reject the traditional claim that "nature abhors a vacuum."

---

[1] Edward Grant, *Much Ado About Nothing: Theories of Space and Vacuum from the Middle Ages to the Scientific Revolution* (Cambridge, Eng., 1981), pp. 86–87.

Because experiment and observation played so small a role in the technical sciences during the Middle Ages, those sciences did not advance significantly. Under such limitations, natural philosophy fared better, since most of the themes it treated were not easily subjected to experiment or mathematization.

INDIANA UNIVERSITY

Agricultural Practice and Garden Design in Renaissance Liguria

by George L. Gorse

When the English royalist adviser, landowner, connoisseur, horticulturist and traveler, John Evelyn (1620–1706), sailed along the coastline of Liguria (fig. 1), first entering Italy on *Le Grand Tour* in October 1644, he described *three* landscape experiences.[1] The first, he rendered in an agitated language that a century later Edmund Burke would theorize as "the sublime."[2] Approaching from Nice, Evelyn saw open before him:

> . . . all this coast (except a little of San Remo) is a high and steep mountainous ground, consisting all of rock-marble, without any grass, tree, or rivage, formidable to look on. A strange object it is, to consider how some poor cottages stand fast on the declivities of these precipices, and by what steps the inhabitants ascend to them. The rock consists of all sorts of the most precious marbles.[3]

At the other extreme were the many harbor enclaves: villages, towns and cities, among these the densely-populated commercial and political center of Genoa (fig. 3), respites Evelyn desperately sought "betwixt those horrid gaps of the mountains" that almost swept away his ship in heavy, stormy seas! To these havens people gathered, building up the protected sides of the mountains above natural harbors, which resembled (for Evelyn and many others) ancient Greek theatres (a metaphor we will return to later).

In between, in a *middle landscape* (fig. 2), Evelyn particularly admired a more tamed, suburban-seaside prospect, filled with noble family villas, that seemed to

---

This article is fondly dedicated to my colleague, Brad Blaine, and my other Scripps colleagues, Sara Adler and John Geerken, with whom I have had such a fruitful relationship over the years in the garden of Medieval and Renaissance studies. I also would like to thank more recent colleagues, Nancy van Deusen and Nancy Bowen, for their inspiration in putting together this well-deserved *Festschrift* in Brad's honor.

[1] *Diary and Correspondence of John Evelyn*, ed. W. Bray, Esq., 4 vols. (London, 1857), 1:83–88 (October 11–19, 1644); J. Stoye, *English Travellers Abroad 1604–1667* (New Haven, 1989), pp. 119ff.

[2] E. Burke (1729–1797), *A philosophical enquiry into the origins of our ideas of the sublime and beautiful*, ed. J. Boulton (London, 1958); J. Hunt, *Garden and Grove: The Italian Renaissance Garden in the English Imagination 1600–1750* (Princeton, 1986), *passim.*

[3] *Diary and Correspondence*, ed. Bray, 1:83.

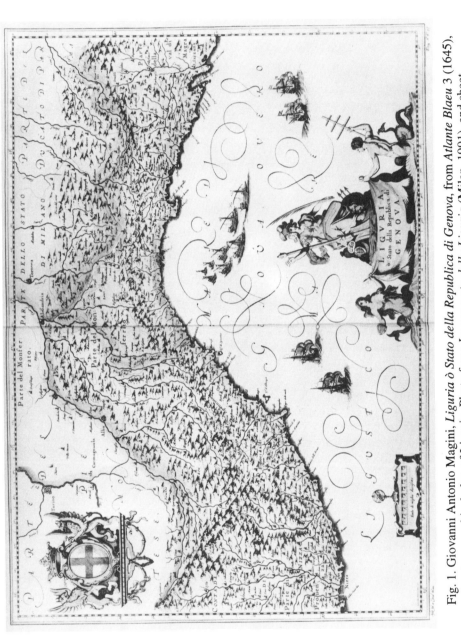

Fig. 1. Giovanni Antonio Magini, *Liguria ò Stato della Republica di Genova*, from *Atlante Blaeu* 3 (1645), topographical plan of Liguria (Milan, 1991), end sheet. Photo from *La scoperta della Liguria*

Fig. 2. Christoph Krieger, "View of Sampierdarena," from J. C. Volckamer, *Nürnbergische Hesperides* (Nürnberg, 1708). Photo from *La scoperta della Liguria* (Milan, 1991), fig. 23.

Fig. 3. Cristoforo de Grassi, *View of Genoa*, 1597, showing city in 1481. Oil on canvas, Civico Museo Navale, Genoa, n. 3486. Photo: Direzione Belle Arti, Genoa.

him "like a continuous city" or "a city up against the city," imitating classical Roman of urbane villa societies along the Bay of Naples or "Bay of Luxury."[4]

> And now [Evelyn exclaimed], as we were weary with pumping and laving out the water, almost sinking, it pleased God on the sudden to appease the wind, and with much ado and great peril we recovered the shore, which we now kept in view within half a league in sight of those pleasant villas, and within scent of those fragrant orchards which are on this coast, full of princely retirements for the sumptuousness of their buildings, and nobleness of the plantations, especially those at St. Pietro d'Arena [just beyond the Lantern and harbor entrance to Genoa, figs. 2–3]; from whence, the wind blowing as it did, might perfectly be smelt the peculiar joys of Italy in the perfumes of orange, citron, and jasmine flowers, for divers leagues seaward.[5]

Notably, Evelyn and other travel accounts, as well as the idealized bird's-eye views meant for the European cultural pilgrim and collector of the seventeenth and eighteenth centuries, largely leave out the technological functions and work dimension of these scenic vistas, concentrating instead on the leisure aspect, conspicuous consumption and display of aristocratic patrons, like themselves, the European travelers.[6] For instance, Evelyn and the *vedute* omit the messy shipbuilding activities concentrated along the beach of Sampierdarena (fig. 2), the center of Genoese shipbuilding, stationed, appropriately enough, right below the suburban seaside retreats of the Genoese merchant nobility.[7] These aristocratic *ville marittime* overlooked their patron's commercial naval interests and the harbor approach, enclosing land and sea with ascending prospects, formal garden terraces and hillside *boschi* (woods) for entertainment and representation,

---

4 Ibid., 1:84ff.; cf. J. d'Arms, *Romans on the Bay of Naples: A Social and Cultural Study of the Villas and Their Owners from 150 B.C. to A.D. 400* (Cambridge, Mass., 1970); esp. Pliny the Younger's description of his seaside villa at Laurentinum (*The Letters of Pliny*, trans. W. Melmoth, Loeb Classical Library [London, 1915], 2.17.24–29): "The sea-front gains much from the pleasing variety of the houses built either in groups or far apart; from the sea or shore these look like a number of cities."

5 *Diary and Correspondence*, ed. Bray, 1:84.

6 For an overview of this pleasure villa construction and aristocratic display, see G. Doria, "Investimento della nobiltà genovese nell'edilizia di prestigio (1530–1630)," *Studi Storici* 27 (1986), 5–55; R. Goldthwaite, *Wealth and the Demand for Art in Italy 1300–1600* (Baltimore, 1993).

7 The largely private shipbuilding in Genoa and Sampierdarena are discussed in J. Heers, *Gênes au XVe Siècle, Activité économique et problèmes sociaux* (Paris, 1961), pp. 267–320 and *passim.* Cf. the public character of Venetian shipbuilding in F. Lane, *Venice: A Maritime Republic* (Baltimore, 1973).

as well as agricultural production.[8] A highly organized, suburban villa-city stretched out from the port and sumptuous palace-city above Genoa, symbolically protected by communal insignias, Christian saints, and classical *genii* (spirit or victory figures) in contemporary civic iconography (figs. 2–3, 10).[9]

What an introduction to Italy! Evelyn unites a Virgilian landscape poetics with Petrarch's Christian redemptive Nature, in harmony with earlier panegyrics to *"Genova La Superba."*[10] From dark into light, like an epiphany from God, the peninsula first revealed itself as a paradisal garden to Evelyn and other northern elite sea voyagers, gazing upon the many coastal pleasure villas of the Genoese nobility, amidst the odiferous terraced gardens and orchards! This paper excavates Evelyn's middle landscape, establishing the close rapport between these garden villas and agricultural practice, not recorded but required for the visual impact of the Genoese villa style, this most powerful urban-landscape image of *"buon governo"* by local noble family groups. In particular, contour terracing (fig. 4) carried out by *contadini*, peasant men and women, on agricultural estates for centuries, resulted in what contemporaries called *"una terra fatta a scalini"* ("a land made into steps"), fundamental to the larger land-use patterns in Liguria from Roman times to the eleventh- and twelfth-century agricultural revolution which contributed to the rise of late medieval Italian communes.[11]

These extensive terracing systems, still visible and partially used today (although diminishing with land abandonment and urban sprawl), bridged town and country within this rugged, mountain-sea environment. They structured the land and made arable a very difficult terrain, while providing a base for prestige building.

Underlying the sweeping, triadic landscape experience of Evelyn and other foreign visitors was the complex patchwork quilt of conflicting jurisdictions and territorial powers which made up the Ligurian coastline and mountainous interior region right through the sixteenth and seventeenth centuries (fig. 5).[12]

---

[8] For an overview of this villa tradition, see E. De Negri, et al., *Catalogo delle ville genovesi* (Borgo San Dalmazzo, 1967); L. Magnani, *Il tempio di Venere: Giardino e villa nella cultura genovese* (Genoa, 1987).

[9] For this tradition of civic iconography, see E. Poleggi, *Iconografia di Genova e delle riviere* (Genoa, 1977), *passim.*

[10] For this earlier rhetorical tradition, see G. Petti Balbi, *Genova medievale, vista dai contemporanei* (Genoa, 1978), esp. Petrarch's encomia to Genoa in 1352 and 1358, pp. 76–83; F. Petrarca, *Le familiari*, ed. V. Rossi, 4 vols. (Florence, 1937), 14.5.23–7; F. Petrarca, *Itinerarium Siriacum*, Opera (Basilea, 1554), pp. 618–9.

[11] For this vernacular agricultural tradition, fundamental to villa and garden design, see E. Poleggi and P. Cevini, *Le città nella storia d'Italia: Genova* (Bari, 1981), pp. 15–21 and *passim*; G. Spalla, *L'architettura popolare in Italia: Liguria* (Bari, 1984); A. Girani and C. Galletti, *Una terra fatta a scalini* (Genoa, 1991).

[12] Heers, *Gênes au XVe Siècle*, Parts 1 and 3; M. Quaini, ed., *La conoscenza del territorio ligure fra medio evo ed età moderna* (Genoa, 1981).

Fig. 4. Example of terracing in Liguria (Cinque Terre). Photo: *La scoperta della Liguria* (Milan, 1991), fig. 85.

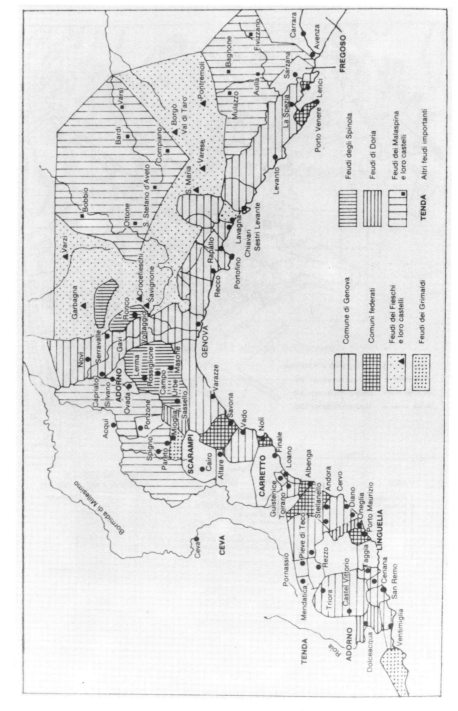

Fig. 5. Regional Map of Liguria, showing family *feudi*. Photo: J. Heers, *Genova nel'400* (Milan, 1991), p. 380.

Beyond the communal territory directly under the control of Genoa, these included the federated or allied communes, most notably Savona, for the most part under Genoa's sway, but always competing and rebellious in their local politics, as well as the old noble family feudatories (often bestowed with Lombard and Hohenstaufen imperial titles and privileges) that controlled much of the Ligurian coastline and inland mountain valley passes. These hereditary *feudi* included the rival Grimaldi, Fieschi, and Malaspina guelph families intermixed with the Adorno, Fregoso, Spinola, and Doria ghibelline families on the eastern and western *riviere*.

During the eleventh and twelfth centuries, these old noble feudal families moved into Genoa and other local communes, taking up commerce, banking, and *condottiere*land and naval commands as competing political factions. They protected themselves in closed family neighborhoods (called *"alberghi"*) which divided the city into tower societies and defensible territories—some close-knit, others loosely bound—just as they had the countryside, where these families retained their agricultural estates and old noble rank in their ancestral seats along the coast.[13] Country estates were particularly valuable assets to urbanized family groups as bases of power on the land for the city, sources of income from landed serfs or rentiers, who worked the terraced fields and local manufacture (often in conflict with their feudal lords), giving protection to these aristocratic lineages during frequent periods of exile and pestilence in Genoa.[14]

The French economic historian, Jacques Heers, spoke of the "two aristocracies" of Genoa as distinct cultures of landed nobles and urban middle class merchants, *signori e mercanti*; but these were actually two sides of the same elite coin.[15] Genoese old noble families often pursued both what we would call landed-military investment and merchant activities, and the Romantic binary construction of "feudal" and "modern" distorts the close interaction of these activities and interests. While the history of medieval Italian communes is the story of commercial elites in cities taking over and controlling agricultural production in their surrounding countrysides (the *contado* or *distretto*), in the case of Genoa the magnates (or landed aristocracy) took over and ran the commercial city. While maintaining and highlighting their noble titles and lands, they struggled to rule over the city and region through territorial-based factions in ever-changing family alliances. Thus, Machiavelli and Guicciardini stressed the

---

[13] Heers, *Gênes au XVe Siècle*, pp. 563–600; E. Grendi, "Profilo storico degli alberghi genovesi," *La repubblica aristocratica dei genovesi* (Bologna, 1987), pp. 49–102.

[14] Witness the many flights of old noble families into exile or from disease in Genoa to their landed estates along the riviera, and the constant conflicts between these families in their urban *alberghi* and coastal *feudi* during the late Middle Ages, recorded in the vivid pages of the Genoese chroniclers; see *Annali genovesi di Caffaro e dei suoi continuatori*, trans. C. Roccatagliata Ceccari and G. Monleone, 7 vols. (Genoa, 1923–29), *passim.*

[15] Heers, *Gênes au XVe Siècle*, Part 3; J. Heers, "The Two Aristocracies—The Case of Genoa," ed. A. Molho, *Social and Economic Foundations of the Italian Renaissance* (New York, 1969), pp. 164–8.

violent and unstable, factious history of Genoa, which involved both *riviere* and their cities.[16]

Fortified towers defined the focal points of major family powers within the city and countryside (figs. 3, 6–7). In a seventeenth-century bird's-eye view of the old noble Doria family stronghold at Dolceacqua on the western riviera, now in ruin, we see the various strata of this feudal territory, redefined through time.[17] Appropriately, this print is entitled *Castrum et Oppidum Dulcis Acquae*, "Castle and Village of Dolceacqua" (or "sweet water"). Above stands the Doria castle fortress, symbol and strength of the local old noble family group with garrison forces; its paired-tower façade and cylindrical keep overwhelm the modest agricultural walled village below, projecting its architectural image out over the surrounding lands and valley approach to the coast.[18]

A triumphal medieval bridge connects the settlement to the opposite shore beyond the Dolceacqua torrent, where a more open and confident Renaissance belvedere banquet pavilion in classical style rests over an enclosed, formal garden with fountains, fruit trees, flowers and herbs; a veritable Garden of Eden for the local lords and their family members. Nearby, a family parish church awaits, giving Christian solace after the pleasures of the flesh celebrated in the formal garden next door. Meanwhile, the peasant population works the terraced fields and vineyards in a "second nature," surrounding the hilltop castle-village and flat triclinium garden, abundant (according to contemporary sources) in fruits, herbs, olive oil, and wine.[19] According to documents of 1447 in the Archivio di Stato of Genoa, we learn that the Doria family suppressed a local attempt to build a mill (*mulino*) on this torrent for fear that the villagers would gain wealth and independence from their local feudal landlords.[20] Thus, here is an instance of conflict between landed interests and technology on the family *feudo*, which one does not often see in *città*.

Returning to Evelyn: even after a frightful entry, his description echoes Petrarch's famous *laudatio* to Genoa of 1352, in which the Italian poet competed (as did Evelyn) with ancient Roman seaside-villa accounts or ekphrasis (dramatic word-image expression) from Cicero, Pliny the Younger, Pausanias, and Statius

---

[16] N. Machiavelli, *Istoria fiorentine*, ed. F. Gaeta (Milan, 1962), 5.6, pp. 335–6; F. Guicciardini, *Storia d'Italia*, ed. F. Catalano, 3 vols. (Rome, 1975), 7.5 (1.215–16).

[17] E. Bernardini, et al., *Dolceacqua dalle origini ai giorni nostri* (Dolceacqua, 1988); P. Stringa, *Castelli in Liguria* (Genoa, 1989), pp. 27–30; B. Durante and A. Eremita, *Guida di Dolceacqua e della Val Nervia* (Cavallermaggiore, 1991).

[18] For this ancient *Herrschaftsarchitektur*, the paired-tower bastion, see K. Swoboda, *Römische und Romanische Paläste, Eine Architekturgeschichtliche Untersuchung* (Vienna, 1969), pp. 77–132.

[19] Quaino, ed., *La conoscenza del territorio*, pp. 79–80 and *passim*.

[20] Heers, *Gênes au XVe Siècle*, p. 530.

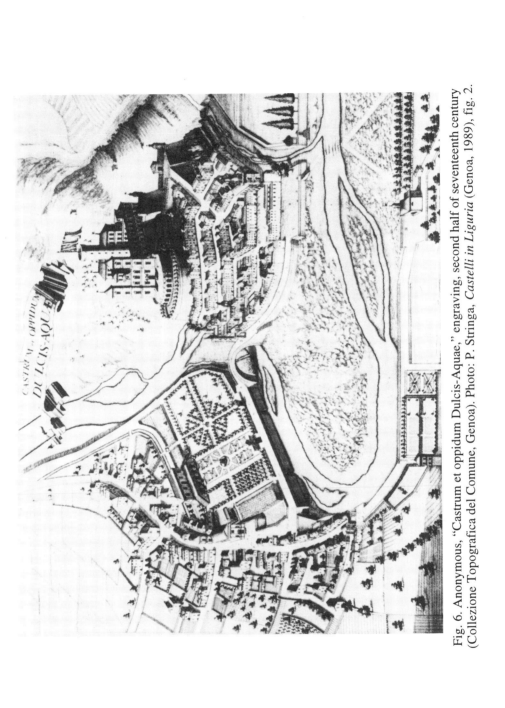

Fig. 6. Anonymous, "Castrum et oppidum Dulcis-Aquae," engraving, second half of seventeenth century (Collezione Topografica del Comune, Genoa). Photo: P. Stringa, *Castelli in Liguria* (Genoa, 1989), fig. 2.

Fig. 7. Dolceacqua Bridge and Doria family castle, fourteenth/fifteenth centuries.
Photo: E. Bernardini, *Dolceacqua dalle origini ai giorni nostri* (Dolceaqua, 1988), p. 19.

(among others).[21] Placing Genoa in an arcadian landscape, Petrarch and other humanists, in turn, influenced Renaissance perspective views of the city, such as Cristoforo de Grassi's large oil painting on canvas (fig. 3), commissioned by the commune to glorify the city in 1597, but based upon an earlier view which commemorated the triumphal return of the Genoese and papal fleet of the Ligurian Pope Sixtus IV della Rovere (reign 1471–84) from naval victory over the Turks at Otranto in 1481.[22] In a classical mode of discourse, Petrarch exclaimed:

> [Genoa] seemed to me not an earthly place, but a celestial abode which the poets place in the Elysian Fields, as the peaks of the hills [rise] with their amenable paths [above] the fertile little valleys and in these valleys [live] happy people. Who would not have gazed with amazement from high [on the surrounding hills] at the towers and palaces, nature vanquished by humans, the rough hills covered by cedars, vineyards and olive groves, the buildings of marble at the foot of the hills, second to none in royalty and enviable to any city?[23]

Outside the city walls, according to the Florentine Giovanni Ridolfi, in 1480:

> there are many houses of citizens that have little land, that is gardens, all enclosed by walls, that the walls appear on the hills here and there like a labyrinth, and to each of these hills they go on their mules to their gardens at the hour of lunch and dinner before returning to the city and their merchant shops.[24]

Merchant class villas, such as the mid-fifteenth-century Villa Tomati (fig. 8) in the Val Polcevera to the west of the city, embodied this tradition of suburban residence and retreat.

The villa's open *belvedere loggia* overlooked an enclosed garden and terraced vineyard below (now destroyed). Its elevated position commanded a panoramic landscape-sea view, visually uniting the triadic landscape, while providing a *pied à terre* in nature over the city, among the many towers, castles and rustic villas of the old nobility—a retreat from the high density of city life, particularly

---

21 See nn. 4 and 10 above; G. Gorse, "Genoese Renaissance Villas: A Typological Introduction," *Journal of Garden History* 3 (1983), 255–80.

22 For a catalogue entry on this major public commission by the magistrates of urban planning, the *Padri del Comune*, see Poleggi, *Iconografia di Genova* 58, p. 112.

23 Gorse, "Genoese Renaissance Villas," p. 255.

24 Ibid., p. 256.

Fig. 8. Villa Tomati in Val Polcevera, c. 1450 with late-fifteenth-century additions (Genoa).

during the hot summer months (*la stagione di villeggiatura*), and protection from the frequentpestilence and family factional strife.[25]

In the early sixteenth century, the Genoese admiral Andrea Doria (1466–1560), leader of the 28 major old noble family clans (*alberghi*) in the foundation of the aristocratic republic of 1528, introduced a Roman classical revival *sea-villa style*, which represented his power and position within the city, the major entry port into Italy and the Mediterranean Empire of the Hapsburg Holy Roman Emperor Charles V (reign 1519–56), whom the admiral served as Mediterranean fleet commander. Between the 1520s and 1540s, Doria bought up a large part of the western trading suburb (*faubourg*) of Fassolo (figs. 3, 9–10), to create a monumental U-shaped harbor villa (figs. 11–12), which opened up the closed, defensive, labyrinthine country houses of the fifteenth century, into a new, perspective unity of space, land and sea.[26]

According to William Thomas' *History of Italy* of 1549, the garden was funda-mental to this new harbor landscape:

> Amongst all other the palace of Andrea Doria, without the gate of St. Thomas [outside the main entrance gate to Genoa on the west], is a notable thing, very fair, sumptuous, and large. And above his house (a thing wonderful) he hath made his slaves to hew out of the hard rocky mountain as much space as hath made *six gardens* one above another, and hath caused so much earth to be carried up as sufficeth for the growth of all manner fruits and herbs, very pleasant to behold [italics mine].[27]

A century later, Evelyn highlighted the Villa Doria in his Genoese description, which in the meantime had been expanded to the sides in the late sixteenth century by the admiral's heir, Giovanni Andrea Doria (1540–1606), as shown in Johann Christoph Volckamer's marvelous published view of 1708 (fig. 11) and a nineteenth-century photograph (fig. 12), taken before the destruction of the upper terraced orchard with its colossal Jupiter statue in a niche, opposite Neptune riding a sea chariot in the lower, seaward, flat formal garden.[28] Evelyn explained:

[25] On the Villa Tomati, see E. Poleggi, "Genova e l'architettura di villa nel secolo XVI," *Bollettino del centro internazionale di studi di architettura Andrea Palladio* 11 (1969), 231–5; *Catalogo delle ville genovesi* 5, pp. 9–61; Gorse, "Genoese Renaissance Villas," pp. 256–7; Magnani, *Il tempio di Venere*, pp. 11–22.

[26] On the Villa Doria, see P. Boccardo, *Andrea Doria e le arti: Committenza e mecenatismo a Genova nel Rinascimento* (Rome, 1989), pp. 25–88, with bibliography.

[27] W. Thomas, *The History of Italy (1549)*, ed. G. Parks (Ithaca, 1963), pp. 106–7. Traffic in slaves for Doria's military business as a naval *condottiere* provided a ready hand source of labor for these large building projects. See E. Grendi, "Andrea Doria, Uomo del Rinascimento," *La repubblica aristocratica dei genovesi* (1987), pp. 139–72.

[28] On Giovanni Andrea Doria's expansion of the villa during the late sixteenth century, see Magnani, *Il tempio di Venere*, pp. 115–24.

Fig. 9. Detail of Villa Doria in Fassolo, from Alessandro Baratta *La famosissima e nobilissima città di Genova con le sue nuove fortificazioni*, 1637. Engraving on 10 plates. Paris, Bibliothèque Nationale, Cabinet des Estampes, Vb. 13. Photo: E. Poleggi, *Iconografia di Genova e delle riviere* (Genoa, 1977), end sheet.

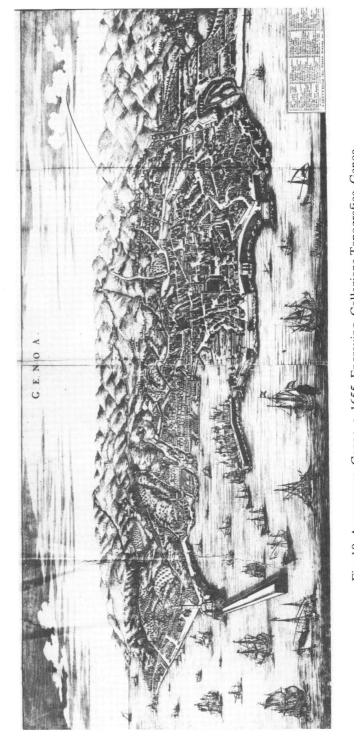

Fig. 10. Anonymous, Genoa, c. 1655. Engraving. Collezione Topografica, Genoa.
Photo: E. Poleggi, *Iconografia di Genova e delle riviere* (Genoa, 1977), fig. 43.

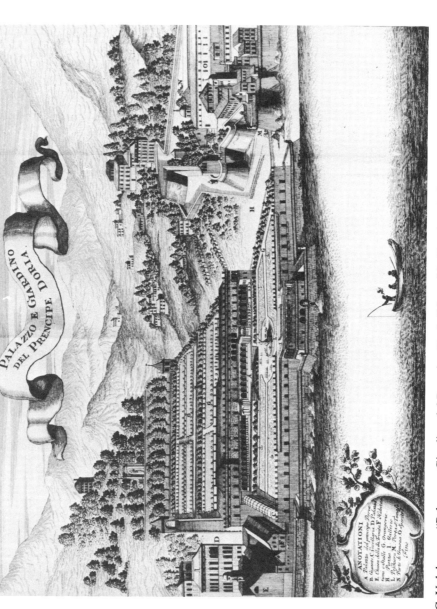

Fig. 11. J. C. Volckamer, "Palazzo e Giardino del Prencipe Doria," from his *Nürnbergische Hesperides die Gründliche Beschreibung der Edlen Citronet—Citronen und Pomeranzen*, Nuremburg, 1708–14, 1:96 (Dumbarton Oaks, Washington, D.C.).

Fig. 12. Villa Doria, late nineteenth century. Photo: Fratelli Alinari.

One of the greatest here for circuit is that of the Prince Doria, *which reaches from the sea to the summit of the mountains* [literally uniting the triadic landscape]. The house is most magnificently built without, nor less gloriously furnished within, having whole tables and bedsteads of massy silver, many of them set with agates, onyxes, carnelians, lazulis, pearls, tourquoises, and other precious stones [like antique luxury villas]. The pictures and statues are innumerable. To this palace belong *three* gardens, the first [below] whereof is beautified with a terrace, supported by pillars of marble: there is a fountain of eagles, and one of *Neptune* [a mythological portrait of Andrea Doria, quelling the seas], with other sea-gods, all of the purest white marble; they stand in a most ample basin of the same stone. At the side of this garden is such an aviary as Sir Francis Bacon describes in his *Sermones fidelium*, or Essays, wherein grow trees of more than two feet diameter, besides cypresses, myrtles, lentiscuses, and other rare shrubs, which serve to nestle and perch all sorts of birds [vying with Varro's ancient Roman aviary], who have air and place enough under their airy canopy, supported with huge iron work, stupendous for its fabric and the charge. The other *two* gardens [above] are full of orange trees, citrons, pomegranates, fountains, grots and statues [like a terraced orchard, combining agricultural production and monumental sculptural decoration]. One of the latter is a colossal *Jupiter* [Father of the Gods on the mountain above the port city], under which is the sepulchre of a beloved dog, for the care of which one of this family received of the King of Spain 500 crowns a year, during the life of that faithful animal. The reservoir of water here is a most admirable piece of art; and so is the grotto over against it.[29]

During the sixteenth and seventeenth centuries, this princely residence and galley-fleet station in front of the city gate and harbor entrance controlled access by land and sea; the villa became the major ceremonial and political reception center into Genoa for visiting Hapsburg monarchs and other European dignitaries.[30] It was a kind of "state within the state," to quote Machiavelli on the earlier Bank of St. George on the medieval harbor front of Genoa, a *suburban* representation of the larger city-state, which existed for the sake of a garden (figs. 9–10).[31] The gardens represented the major civic themes of dominion on *sea* and

---

[29] *Diary and Correspondence*, ed. Bray, 1:86.

[30] G. Gorse, "Between Republic and Empire: Triumphal Entries in Genoa during the Sixteenth Century," *"All the world's a stage . . . ": Art and Pageantry in the Renaissance and Baroque*, ed. B. Wisch and S. Munshower (University Park, Penn., 1990), 1:189–256.

[31] On the Banco di San Giorgio, incorporated in 1407 to assume the public debt of Genoa, see H. Sieveking, "Studio sulle finanze genovesi nel Medioevo e in particolare sulla Casa di S. Giorgio," *Atti della Società Ligure di Storia Patria* 35 (1906), 3–368; L. Grossi Bianchi and E. Poleggi, *Una città portuale del Medioevo: Genova nei secoli X–XVI* (Genoa, 1980), pp. 173, 244, 296–301; G. Rotondi Terminiello, *Palazzo San Giorgio, Guide di Genova* 38 (Genoa, 1977).

land, "*la signoria del mare e della terra*," to cite one Renaissance description of Genoa.[32] The kingdoms of *Neptune and Jupiter* were balanced within a flat floral parterre and ascending orchard order—a formal presentation of the "two natures" in a "*terra fatta a scalini*"[33]—which created a vast harbor-villa, garden-mountain, axial-theatre for the Doria old noble family court at the entrance connection between the Genoese Republic and the Hapsburg Mediterranean Empire. It was a place for pleasure retreat and politics, around which the entire city was reorganized (much like contemporary central Italian ducal *villa-neighborhoods*, such as the Medici court at the Boboli Gardens, reorganized Florence).[34]

In the mid-sixteenth century, a second phase of the Genoese Renaissance villa development began with the arrival of the Perugian architect, Galeazzo Alessi (1512–72).[35] While this Roman-trained architect was brought to Genoa by Cardinal Bartolomeo Sauli's family to build their monumental neighborhood church of Santa Maria Assunta in Carignano, in imitation of Bramante's centralized church of St. Peter's, high atop a hill to the east overlooking the port (fig. 10), Alessi also received commissions from other old noble families of the merchant aristocracy to design a closed version of the axial Via Giulia in Rome, a *Strada Nuova* for their urban palaces above the medieval center;[36] as well as a classical dome and choir for the medieval Cathedral; a triumphal entry gate on the medieval harbor front; and classical villas for the aristocratic suburbs. Alessi's *Roman* style became the hallmark of the Genoese old noble family republic, and the *siting* of these buildings was crucial to the new *Roman* image of Genoa, setting up a symbolic architecture and topography for the Hapsburg banking emporium with its many family connections to the Eternal City.[37]

In the suburb of Albaro above the *Val Bisagno* to the east of Genoa, balancing the coastal villa suburb of *Sampierdarena* described by Evelyn to the west, Alessi

---

[32] F. Filelfo, "Descrizione di Genova (1448–1450)," in *Genova medievale vista dai contemporanei*, ed. G. Petti Balbi, p. 131.

[33] Cf. the contemporary Villa Lante at Bagnaia and theoretical discussions of the "two natures" in garden representation; see C. Lazzaro, *The Italian Renaissance Garden: From the Conventions of Planting, Design, and Ornament to the Grand Gardens of Sixteenth-century Central Italy* (New Haven, 1990), *passim.*

[34] M. Fagiolo, "Effimero e giardino: Il teatro della città e il teatro della natura," *Il potere e lo spazio: La scena del principe* (Florence, 1980), pp. 31–54; R. Burr Litchfield, *Emergence of a Bureaucracy: The Florentine Patricians 1530–1790* (Princeton, 1986).

[35] C. Maltesi, et al., *Galeazzo Alessi e l'architettura del Cinquecento*, Atti del convegno internazionale di studi, Genova, 16–20 Aprile 1974 (Genoa, 1975), pp. 289–448.

[36] Cf. L. Salerno, et al., *Via Giulia, una utopia urbanistica del 500* (Rome, 1973); M. Labò, "Strada Nuova (più che una *strada*, un *quartiere*)," *Scritti di Storia dell'Arte in onore di Lionello Venturi*, 2 vols. (Rome, 1956), 1:403–10; E. Poleggi, *Strada Nuova, una lottizzazione del Cinquecento a Genova* (Genoa, 1968); F. Caraceni, *A Renaissance Street: Via Garibaldi in Genoa* (Genoa, 1993).

[37] On the special relationship between Genoa, the Hapsburg Empire, and Rome during the sixteenth century, see R. Lopez, "Predominio economic dei Genovesi nella monarchia spagnola," *Giornale storico e letteratario della Liguria* 12 (1936), 65–74; C. Costantini, *La Repubblica di Genova* (Turin, 1986), *passim.*

built a Roman style villa for the Genoese merchant-banker, Luca Giustiniani, a prominent member of the old noble elite, in 1548–50. (Fig. 13 shows Robert Reinhardt's 1886 photograph of the villa, which still records the remnants of the formal axial garden before its conversion to an English picturesque landscape park after a World War II restoration.)[38]

Inspired by mid-sixteenth-century ideas about ancient Vitruvian houses and contemporary Roman palaces and villas, the Villa Giustiniani-Cambiaso initiated a new style, characterized by a dynamic, symmetrical, classical, *cubic block*, set on a hill in axial relationship to a formal garden and a sea view down the long valley.[39] Evelyn focused his description of Genoa on the classical idea of a dramatic urban scenography with its key element of *social visibility* or *representation*. He said:

> The city [of Genoa] is built in the hollow or bosom of a mountain, whose ascent is very steep, high, and rocky, so that, from the Lantern and Mole to the hill, it represents the shape of a *theatre*; the streets and buildings so ranged one above another, as our seats are in *play-houses*; but, from their materials, beauty, and structure, never was an artificial scene more beautiful to the eye, nor is any place, for the size of it, so full of well-designed and stately palaces, as may be easily concluded by that rare book in a large folio which the great virtuoso and painter, [Peter] Paul Rubens, has published, though it contains only one street [the *Strada Nuova*] and two or three churches [italics mine].[40]

Actually, Rubens published Alessi's Villa Giustiniani-Cambiaso in façade elevation and plan (figs. 14 and 14a) as an image of "nobility" for a northern

---

[38] *Catalogo delle ville genovesi*, pp. 415–23; *Galeazzo Alessi e l'architettura del Cinquecento, passim*; Magnani, *Il tempio di Venere*, pp. 59–64.

[39] See discussion of the Genoese palace and villa style in M. Labò, "L'architettura dei palazzi genovesi," *Lo Spettatore: Rivista di Lettere, Arti, Scienze, Politica* 1 (1922), 2:141–51; M. Labò, "Le ville genovesi," *Emporium* 87 (1938), 131–44.

[40] *Diary and Correspondence*, ed. Bray, 1:84–5.

Fig. 13. Galeazzo Alessi, Villa of Luca Giustiniani, 1548–50, later Cambiaso family.
Photo: R. Reinhardt, *Palast-Architektur von Ober-Italien und Toscana vom XV. bis XVII. Jahrhundert—Genua* (Berlin, 1886), Taf. 26.

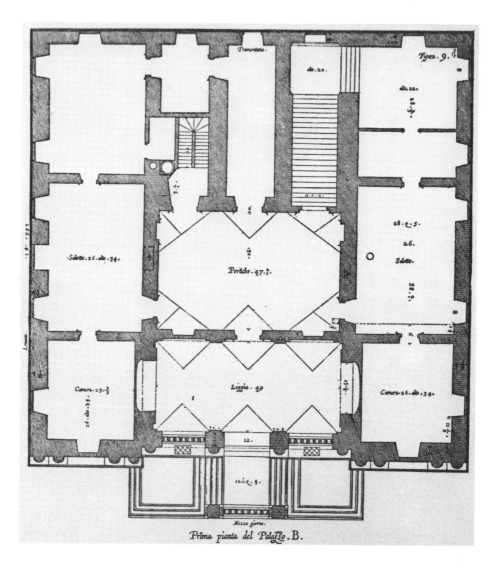

Figs. 14 and 14.a. Peter Paul Rubens, Plan and Façade Elevation of "Palazzo B," from his *Palazzi di Genova* (Antwerp, 1622). Villa of Luca Giustiniani in Albaro, later Cambiaso family ownership. Photo: M. Labò, *I Palazzi di Genova di Pietro Paolo Rubens e altri scritti d'architettura* (Genoa, 1970), pp. 48–9.

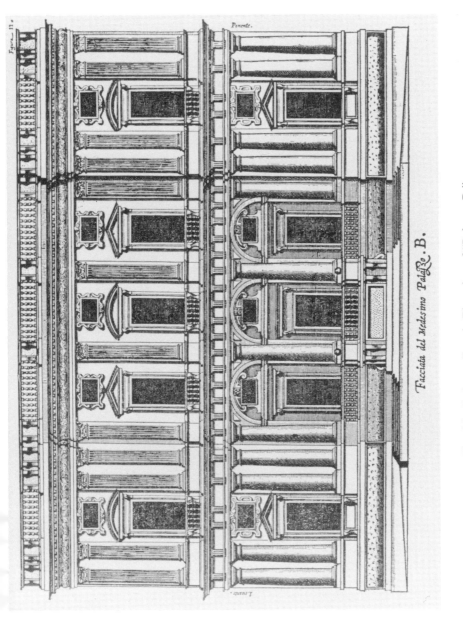

Fig. 14a. Rubens, Façade Elevation of "Palazzo B."

European clientele, as announced in his preface, in homage to the Genoese old noble style.[41]

At the top of the urban theatre, the Alessian villa was like a "sky box" in a modern stadium (figs. 10, 13, 16–17), a conspicuous manifestation of aristocratic family presence and prestige.[42] The Vitruvian Doric and Ionic orders with central belvedere loggia and symmetrical plan with flanking corner towers—an authoritative continuation of the late medieval family *castrum* (figs. 6–7, 8, 11–12, 13, 16–17)—organized not only the façade of the villa but also the interior space, raised up on a platform, oriented outward and ordering the central garden axis, extending down the hill to a sea view (figs. 10, 13). While inland, Alessi's "platform villa" (as it has been called by Donna Salzer) connected the suburban pleasure retreat to the eastern harbor approach to Genoa along a visual axis, relating the patron to a classical tradition of coastal villa scenography and the actual source of wealth—the sea. Classical image framed and redefined the city and urban elite: this type of "conspicuous consumption" was *not* wasted investment, in disagreement with Roberto Lopez's classic thesis on artistic patronage as a response to economic "crisis" and the lack of "real" investment.[43] Representation *is* functional and it *is* real in creating (and recreating) social orders.

In another project, Alessi *simultaneously* designed an urban palace (illustrated in Rubens' "Palazzo A" façade, fig. 15) near the eastern end of the *Strada Nuova* and a suburban villa (figs. 16–17) on the nearby hill above the *Porta Santa Caterina* for the extraordinarily rich Genoese alum merchant-banker, Tobia

---

[41] M. Labò, *I Palazzi di Genova di Pietro Paolo Rubens e altri scritti d'architettura* (Rome, 1970), pp. 1–10, 48–53; G. Biavati, et al., *Rubens e Genova, Catalogo della Mostra* (Genoa, 1977), *passim*. In his preface, Rubens articulated a classical theory of history, architecture, and nobility as central reforms for his deluxe volume of Genoese palaces for northern patrons: "Vediamo che in queste parti (cioè in Fiandra) si va poco a poco invecchiando et abolendo la maniera d'Architettura che si chiama Barbara e Gothica; et che alcuni bellissimi ingegni introducono la vera simmetria di quella, conforme le regole de gli antichi Greci e Romani, con grandissimo splendore et ornamento della patria. . . . Mi è parso dunque, prosegue, di fare una opera meritoria verso il ben publico di tutte le Provincie Oltramontane, producendo in luce li dissegni da me raccolti nella mia peregrinatione Italica, d'alcuni Palazzi della superba città di Genova. Perchè si come quella Repubblica è propria de Gentilhuomini, così le loro fabriche sono bellissime e commodissime, a proportione più tosto de famigle benchè numerose di Gentilhuomini particolari, che di una Corte di un Principe assoluto" (from Labò, *I Palazzi di Genova di Pietro Paolo Rubens*, p. 25).

[42] Cf. aristocratic representation in ancient Roman theatres in P. Zanker, "Status and Applause: The Theater as Meeting Place of Princeps and People," *The Power of Images in the Age of Augustus* (Ann Arbor, Mich. 1988), pp. 147–56.

[43] R. Lopez, "Hard Times and Investment in Culture," *The Renaissance: A Symposium* (New York, 1953), pp. 19–32. For previous discussion and criticism of the Lopez crisis theory of artistic patronage, see Goldthwaite, *Wealth and the Demand for Art*, p. 14ff.; and Doria, "Investimenti."

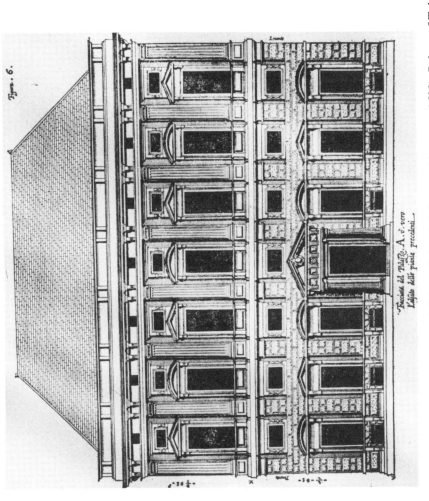

Fig. 15. Peter Paul Rubens, "Palazzo A," from *Palazzi di Genova* (Antwerp, 1622). Palace of Tobia
Pallavicino by Galeazzo Alessi, 1550–55. Later Carrega Cataldi family, Via Garibaldi 4.
Photo: M. Labò, *I Palazzi di Genova di Pietro Paolo Rubens e altri scritti d'architettura* (Genoa, 1970), p. 40.

Veduta dal Gauthier.

Fig. 16. Villa of Tobia Pallavicino, called "delle Peschiere" ("of the fish ponds"), c. 1555–60. Engraved view from M. P. Gauthier, *Les plus beaux édifices de la ville de Gênes et de ses environs*, 2 vols. (Paris, 1818, 1830–32). Photo: *Catalogo delle ville genovesi* (Borgo San Damazzo, 1967), p. 101.

Fig. 17. Villa Pallavicino "delle Peschiere," 1555–60. Photo: *Catalogo delle ville genovesi* (Borgo San Damazzo, 1967), p. 100.

Pallavicino—an investment of about 80,000 gold *scudi*![44] Pallavicino was notorious for his seignorial patronage spending, and this played a key part in his urban *and* suburban self-promotion within the patriciate and Hapsburg clientele society. Standing prominently on the *Strada Nuova*, the monumental palace of Tobia Pallavicino helped to define the corporate character and the *antiquity* of the Genoese old noble families who commissioned these conspicuous, luxury palaces in the Alessian *antique* style on this enclosed *strada di rappresentazione* atop the medieval city and Renaissance republic. Alessi's self-consciously classical Roman style was symbolic of the *new city* and the *old noble families*, who led the oligarchic republic. The closed block form with central Doric temple-front portal and family crest, rustic-fortress base, and displayed Ionic order on the *piano nobile* of the urban palace-street (fig. 15) were expanded out into a formal garden at the open loggia-villa with corner towers (figs. 16–17), where a series of monumental cubic, stepped terraces were excavated, descending and connecting the villa to the city and harbor theatre.

At the Villa Pallavicino "delle Peschiere" of 1558–60, shown in the French Martin Pierre Gauthier's 1818 view (fig. 16), Alessi enlarged the platform villa to such a grand scale that contemporaries compared it to the walls of a fortified city—a city above the city. Climbing the hill from the *Strada Nuova* family palace, through the city gate of *Santa Caterina* to the villa entry, a Doric portal and enclosed cypress *allée*, the narrow bent-axis approach suddenly opened up to an axial series of belvedere terraces at different levels of the garden with opposing views over the urban landscape. The *garden* connected city and suburb, palace and villa, in an ascending perspectival, panoramic order, much like the earlier Villa Doria on the harbor front (figs. 9–12). At the center of the garden was a monumental Doric grotto with family crest and classical busts, comparing the old noble family to ancient Roman ancestors, that served as a triumphal vestibule or entryway from the city into the villa: a much-needed, cool garden *ricetto* after the arduous climb, connected in a stepped Vitruvian order to the Ionic and Corinthian pilaster façade above.

During the second half of the sixteenth century, the period exactly corresponding to Alessian construction on the *Strada Nuova* and other major projects (c. 1550–85), including the villas, a communal office and a set of formal *Books on the Ceremonies of the Republic of Genoa* (*Libri dei Ceremoniali della Repubblica di Genova*) were drawn up.[45] These included a list ("*La Sorte*" or hierarchical order) of official urban and suburban-country residences for the courtly protocol of reception, entertainment and temporary residence of visiting monarchs, prelates, and dignitaries, ranked in three orders according to their social positions: *prima classe* for popes, rulers, and cardinal legates; *seconda*

---

[44] Labò, *I Palazzi di Genova*, pp. 40–47; *Catalogo delle ville genovesi*, pp. 100–17; Doria, "Investimenti," pp. 16, 29; Magnani, *Il tempio di Venere*, pp. 68–4.

[45] L. Volpicelli, "I libri dei ceremoniali della Repubblica di Genova," *Società Ligure di Storia Patria* 49, fasc. 11 (Genoa, 1921). Discussed by E. Poleggi, "Un documento di cultura abitativa," *Rubens e Genova*, pp. 85–122; Doria, "Investimenti," pp. 9–12.

classe for cardinals and nobles ("*signori di qualità*"); and *terza classe* for "*signori di minor qualità*" (merchants and ambassadors, etc.)

Alessi's Ville Giustiniani-Cambiaso and Pallavicino "delle Peschiere" both ranked prominently on the list of "*prima-classe*" ceremonial centers, in which the garden played a major part in projecting the *Roman* image of formal entry and pleasure retreat, overlooking the urban theater: *la nuova porta d'Italia*. Perhaps, in conclusion, the prominence of Alessi's Roman-style "platform villas" as the model for Genoese noble villas during the next century is explained, in part, by a passage from Giambattista Confalonieri's *Description of Genoa* of 1592:

> Also outside the Portal of Santa Caterina there is a palace of the Pallavicini; the most magnificent of all: there is not one that is better situated, since it dominates all the city, built in such a place, that in the event of a disturbance, this is the first [vantage point] to spy it ("*spianarlo*"). Of note are the grandness and magnificence of the rooms with very famous pictures by a most beautiful hand [the Genoese painter, Luca Cambiaso], and the fishponds and the fountain [in the garden] so artificially made, that neither in Rome nor in Tivoli is there a similar thing.[46]

From agricultural practice to monumental Roman villa style, the Genoese suburban residence contributed to a monumental tradition of court reception, entertainment, production and state protocol in early modern Europe, which still can be seen at the White House in Washington, D. C.

POMONA AND SCRIPPS COLLEGES

---

[46] Quaini, ed., *La conoscenza del territorio ligure fra medio evo ed età moderna*, pp. 130–1.

Technology Illustrated: Problems of Sources

by Richard W. Unger

Brad Blaine has relied heavily on works of medieval artists and has effectively exploited what they did to argue about the evolution of the water mill. Drawings which he has found after a long campaign of research have proven invaluable for his discussions of the technology of milling. He continues to take opportunities to search various parts of Europe for any and all pictures of water mills, trying to supplement and expand the known corpus of such illustrations. In such an exercise he is not alone. For everyone working in the history of medieval technology, the products of artists are a central and even critical source. The power and concision of visual evidence give illustration a central role in documenting and understanding the nature of technology. There are other compelling reasons to exploit such evidence.

Though historians of technology have long known the value of illustrations to their research and long exploited the products of artists, the work of those historians of technology has not been in art history. Their work was never intended to advance the art historical project, and if it did, the gain was purely incidental. The skills of the art historian are rarely in the collection of talents that historians of technology bring to their task. Instead they have in the past and will continue to rely on the work of art historians. The incidental nature of contributions to art history from the use of illustration in the study of the history of technology will remain universally true, and the traffic will continue, equally incidentally, in the opposite direction.

As with the study of any history, there can be no work without sources. The reliance on illustrations for work on medieval technology comes to a significant degree from the paucity of alternatives. Writers in the Middle Ages rarely talked about technology. When they did, the products were rarely reliable. The tendency was to draw on classical texts, merely reproducing what they said. The early twelfth-century treatise on painting, glassmaking and metalwork of a monk named Theophilus is a rare and extremely valuable work on technology. It is, though, unique and suffers from being both narrow in scope and little more than a recipe book.[1] If works like those of Theophilus are the best that can be hoped for from the High Middle Ages, then there is a serious gap in the sources available. That gap forces historians to rely more heavily on illustrations. Pictures known for artistic or other reasons often contain information about technology. The illustration was perhaps incidental for the artist or the observer of the day, but such information forms essential data for the historian of technology.

---

[1] Theophilus, *On Divers Arts: The Foremost Medieval Treatise on Painting, Glassmaking and Metalwork*, trans., intro. and notes J. G. Hawthorne and C. S. Smith (Chicago, 1963); Bertrand Gille, "Prolégomènes à une histoire des techniques," in *Histoire des techniques*, ed. B. Gille (Paris, 1978), pp. 80–81.

The use of illustrations by anyone, but especially by historians of technology, raises serious questions. The questions are not just the standard ones common to the use of any historical source. Issues of accuracy and reliability are certainly important ones and ought not to be dispensed with lightly. Dating the illustration, identifying the location being depicted, isolating the influences on the artist are often extremely difficult jobs. Style is not always a dependable indicator of when a work was produced, but that is often the best one available. Rarely were works dated. Artists moved, and so did works of art. Though sculpture tended to stay put because free-standing sculpture was rare in the Middle Ages, works in virtually all other media were objects of trade and so were moved, and in some cases moved over long distances. Displaced physically and temporally, it is difficult for art historians to be specific about many aspects of a work, and that in turn undermines the value and force of the illustration as a source for the historian of technology.

There is more to reliance on illustrations, however, than just making sure that the historian has the right date and place for the work of art. There are questions about the purpose of the artist's work, about what the artist intended. There is a possibility, and a strong possibility, that the purpose may well have affected what he saw and also affected what he showed. The problem is made acute and indeed critical to the study of the history of technology because the quality, but even more important, the character, nature, purpose and the entire conception of illustration of techniques changed in late medieval and early modern Europe.

In using illustrations the issue is more than simply asking about accuracy. It is also more than the art historian's study of iconology. The history of technology demands much more than finding out what an object represents. Ideally, the illustration should yield information about the technology itself. It should also yield information about the vision of technology conceived and understood by the artist. It should also inform how technology created or effected social relations and social change. In trying to extract that broad and complex range of information, for the historian of technology it is as much a matter of what the illustrator leaves in as what the illustrator leaves out which is of interest and of value. Few if any would dispute the old proverb that a picture is worth 10,000 words, but, of course, the problem is trying to decide which collection of 10,000 words is best.

To flirt with oversimplification, in the Middle Ages illustration of technology was incidental. There was virtually no conscious effort on the part of any artist to show how things worked. The earliest possible candidate in the West is the notebook done by Villard de Honnecourt in c. 1235. It contains a number of sketches of various devices and methods which turn up there for the first time, such as gadgets for controlling clocks and the use of a spring in machine design and the water-driven sawmill. The sketches indicate that such equipment was known and suggests that the technologies associated with them were in use for some time before the author jotted them down.[2] Villard's work was an isolated

---

[2] Gille, "Prolégomènes à une Histoire," p. 95; Lynn T. White, Jr., *Medieval Technology and Social Change* (Oxford, 1962), pp. 82, 118, 120.

one. Other how-to manuals were few and far between, and the ones that did exist had little if any illustration. Artists set out to show Bible stories and the lives of saints. They had no intention whatsoever of showing methods of doing work.[3]

Written works on technology became more common in the fifteenth century and in the Renaissance. Along with them came more illustration. Works like the *Bellifortis* of Konrad Kyeser, who was born in 1366, and the later *Bellicorum instrumentorum liber* of Giovanni Fontana were part of a body of material on warfare and the machines of warfare, as was Vannoccio Biringuccio's *Pirotechnica* of 1540. Kyeser's book or sketchbook was truly unique. It dates from around 1405 and was never published. The work contained multiple drawings of ways to apply simple devices to problems. The books of Fontana and Biringuccio reached a wider audience because they were produced on that new device of the fifteenth century, the printing press. The desire to use that new technology to make available classical texts also contributed to the publication of books on technology. Vitruvius was one author who, early in the sixteenth century, got published. The printed works, whether classical or contemporary often relying on classical precedents, reached a wider audience not only because of the form and format but also because they tried to be more comprehensive. The tendency toward greater range became more pronounced as the sixteenth century wore on. Jacques Besson's *Théâtre des instruments mathematiques et mechaniques* of 1569, reprinted in 1587 (fig. 1), was followed in 1588 by the much more elaborate *Le diverse et artificiose machine* by Agostino Ramelli (1531–1590) with its more detailed, complex and precise drawings (fig. 2). These works tended to be pictures with commentary, something like notebooks but still a great step beyond the sketchbooks of the Middle Ages.[4] The crowning example of the new type of work was the lengthy treatise on mining and metallurgy by a physician from the mining town of Joachimstal, Dr. Georg Bauer, who Latinised his name to Agricola. His *De re metallica* of 1556 went beyond simply sketches or sketches and commentary. Instead, it was a treatise on the technology itself, and to explain that technology he included abundant illustrations, carefully drawn (fig. 3). The depictions combined, not always successfully, realistic and diagrammatic representation and were clearly intended to be explanatory.[5]

Bauer, along with others in the sixteenth century, started certain protocols and conventions about the depiction of technology which would become standard in the coming centuries. The use of cutaway diagrams and identifying letters for various features which are then described in captions or in the text, making it

---

[3] Vaclav Husa, et al., *Homo Faber* (Prague, 1967), p. 12.

[4] Bert S. Hall, "Der meister sol auch kennen schreiben und lesen: Writings about Technology ca. 1440–ca. 1600 A.D. and Their Cultural Implications," in *Early Technologies*, ed. Denise Schmandt-Besserat (Malibu, 1979), pp. 49–53; White, *Medieval Technology*, pp. 128, 175.

[5] G. Agricola, *De re metallica*, trans. H. C. and L. H. Hoover (New York, 1950).

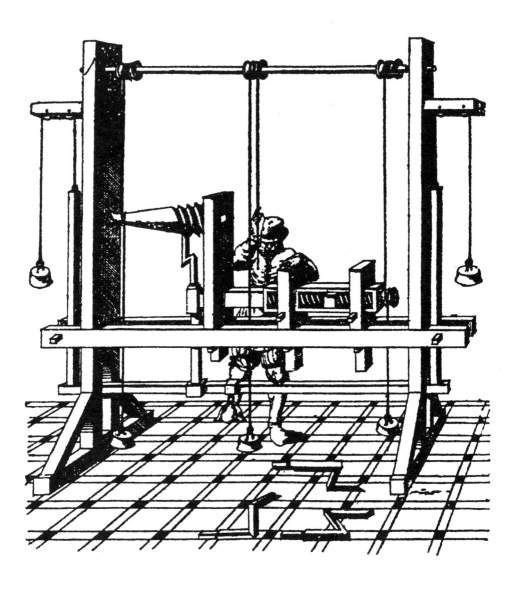

Fig. 1. The screw-cutting lathe in Jacques Besson's *Théâtre des instruments mathematiques et mechaniques*, 1569. From Charles Singer, et al., eds., *A History of Technology*, 5 vols. (Oxford, 1957), 3:334.

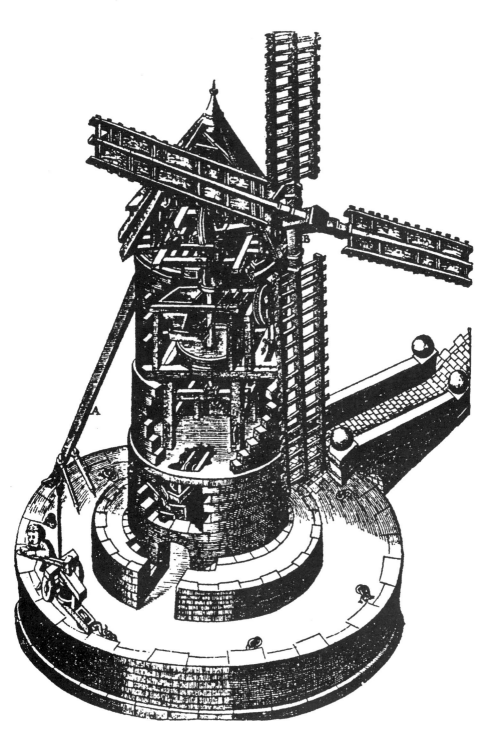

Fig. 2. Tower windmill for grinding grain, from *Le diverse et artificiose machine* by Agostino Ramelli (1531–1590). From Singer, *A History of Technology*, 3:90.

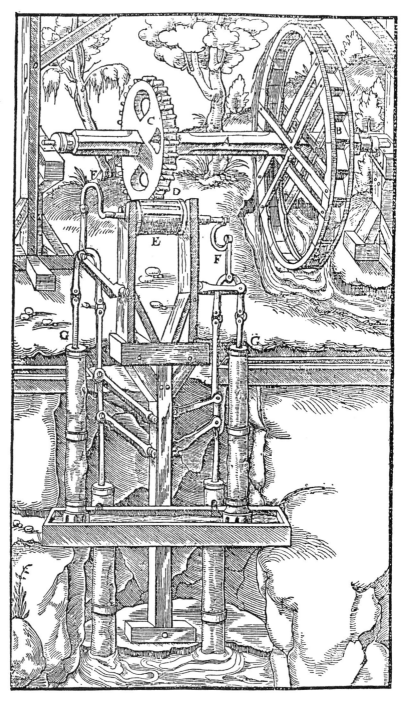

Fig. 3. Water-driven pumps for draining mines from G. Agricola, *De re metallica* of 1556. From G. Agricola, *De re metallica*, trans. H. C. and L. H. Hoover (New York, 1950), p. 189.

possible to associate words precisely with specific items in the illustrations, make extracting information from such works much easier for modern eyes. Authors were experimenting with different forms of illustration of technology and, in the process, found superior methods. They also established traditions which became entrenched in the following two hundred years.

An interest in art and science and a need for guidance in what were increasingly complex technologies combined to create a market for illustrated handbooks. In addition, in the fifteenth and even more in the sixteenth centuries, showing work scenes, a tradition among artists which dated back to the thirteenth century, became more common. That practice led, by the height of the Renaissance, to the production of books of trades with a drawing of different craftsmen each at work at his, or sometimes her, trade. The practice became so normal that by the late sixteenth century depictions of workers appeared as incidental illustration on items as common as, for example, playing cards.[6] In addition, painters came to include work and workers as subjects in the increasingly popular landscapes and genre paintings. In the sixteenth and seventeenth centuries in the Low Countries, a number of artists showed mining and metallurgy. The tendency extended even to Italy where, for example, around 1570 Giorgio Vasari ordered a set of murals on the work of various craftsmen, including dyers and a number of other skilled artisans, for the walls of Florence's Palazzo Vecchio.[7] The intensity and drama characteristic of Renaissance Italian painting was in no way diminished in the depictions, but they still did show workers employing varied techniques. The surroundings in the workshops might owe more to classical models than to contemporary practice, but the tools and ways men were shown working reflected sixteenth-century Tuscany.

In the closing years of the seventeenth and in the eighteenth centuries, a more precise and systematic approach to the illustration of techniques began to emerge. The end result by the 1750s could be called, with justice, industrial design. That great compendium of human knowledge, the *Encyclopédie*, composed by the French *philosophes* and for which Denis Diderot deserves much of the credit, was filled with sumptuous illustrations, marking the completion of the transformation toward a recognizably different way of showing machines and how people made them work (fig. 4). The use of relatively inexpensive copper engraving and the now fully-developed skills of perspective made possible the depiction of machines with accuracy and seeming simplicity. The volumes of plates in the *Encyclopédie*, which included drawings of dozens of devices from carriages to ships to olive presses, demonstrated the now widespread interest in apparatus, an interest in technology *per se* and also the existence of a market for

---

6 Benjamin A. Rifkin, "Introduction," in Jost Amman and Hans Sachs, *The Book of Trades (Ständebuch)* (New York, 1973), pp. xxi, xxxvi, xxxix.

7 F. D. Klingender, *Art and the Industrial Revolution*, ed. and rev. Arthur Elton (London, 1968 [1st ed. 1947]), pp. 57–58.

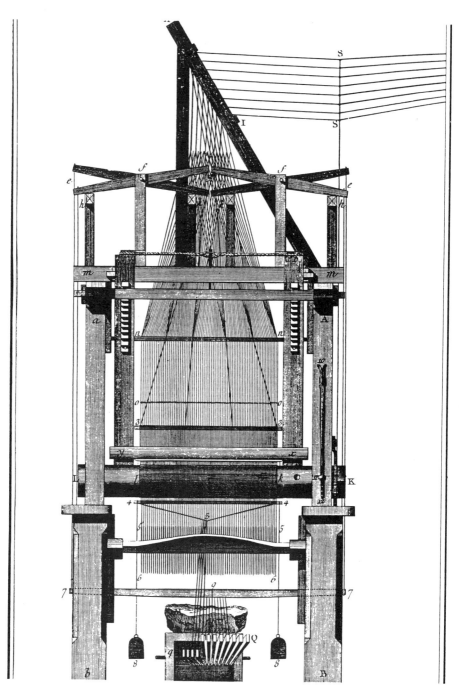

Fig. 4. A silk weaving machine "pour fabriquer les étoffes brochées" from the *Encyclopédie* of 1751–1780. From Roland Barthes, Robert Mauzi and Jean-Pierre Seguin, *L'univers de l'Encyclopédie Diderot et d'Alembert* (Paris, 1964), p. 51.

such illustrations.[8] The artists' work was done with a consciousness of accuracy and a desire for completeness. The men who edited and published such works typically shared a knowledge of and confidence in mathematical formulae to describe the world. They expected the illustrations to reflect, in their purity and precise proportions, exactly such relationships.

With the Industrial Revolution also came depictions, at least by some artists, of workplaces, of machines and their operators at work. Between the sixteenth and second half of the eighteenth century there seems to have been a break among artists in the tradition of producing such illustrations. In the seventeenth century they showed less interest in describing technology in their work than their predecessors or successors. It seems unlikely that the artists were reacting to a change in thinking about technology or about art. Neither resurgent Catholicism nor Protestantism, the principal intellectual and religious trends of the period, condemned labor or the depiction of it. The one obvious exception to the general trend away from showing working people was in Dutch seascapes, where painters, possibly without being conscious of what they were doing, portrayed fishermen and other sailors at times struggling against the elements. There can be no question of the high degree of accuracy in the paintings of those Dutch artists nor of the generally strong Protestant emotions of the emerging state where those artists lived and sold their works. Whatever the reason for the hiatus in the willingness to show people at work, it was after the economy began to expand and also to change character from the mid-eighteenth century, first in England, that some artists even made industry itself the subject of their work. The most obvious and perhaps extreme example was the English artist, Joseph Wright of Derby, who concentrated on the emerging metallurgical industries in the Midlands for the topics of some of his paintings.[9] He was an artist of broad interests and complex views about art and its function.[10] Though both Wright, but more so, the editors of the *Encyclopédie*, thought they were producing realistic depictions of life, in neither case did reality escape their interpretation. On the other hand, both did have an interest in technology and labor, and both granted a significant status to work and how it was done, which their predecessors were less willing to do.

With Wright, as with other artists in the Industrial Revolution, depictions of technology were very different from the now standardized method of showing machines and how they worked. The draftsman had by the eighteenth century been separated from the artist. The draftsman was different because of the use of different techniques and because of his very different purpose. The draftsman was to illustrate technology, not to impart form and beauty, which were the province of the artist. The distinctions between art, science and technology are rather recent. However, such distinctions certainly were being made in

---

8 See Roland Barthes, Robert Mauzi and Jean-Pierre Seguin, *L'univers de l'Encyclopédie Diderot et d'Alembert 1751–1780* (Paris, 1964).

9 Klingender, *Art and the Industrial Revolution*, pp. 51, 55, 59–63.

10 E.g. David Solkin, *Painting for Money: The Visual Arts and the Public Sphere in Eighteenth-Century England* (New Haven and London, 1993), pp. 214–246.

illustrations in the eighteenth century. The word draftsman, meaning someone who makes drawings or designs, was not used in English until 1663. The word draught, meaning to draw, dates at least from the middle of the thirteenth century, but it was more than four hundred years before the language had a noun for the individual who prepared a visual representation of some projected construction. The coinage of the word draftsman presumably reflects the development of a new trade, a new task, and indicates an emerging distinction between artists and illustrators of technology.

The evolution of the approach from the Renaissance through the eighteenth century to the Industrial Revolution turned technical illustration into a matter of showing stages and steps. In the process of producing some good, the use of some technology was reduced to its constituent, or rather what were perceived to be, its constituent parts. The actions of workers were simplified to make them more easily understood and more easily illustrated. That simplification, however, meant the stripping away of the context. Technology was, by the Enlightenment, not shown as part of an organic whole. The breaking of technology and its illustrations down into component parts meant isolation and, in the process, changed the value and the use of illustrations of technology for contemporaries and also for historians of technology.

Building ships demonstrates the long-term evolution in the showing of technology, though it is true that examining the depiction of almost any other technology would glean the same impression. In the early Middle Ages there was no purposeful illustration of shipbuilding. Where artists show a ship under construction, it is to show Noah, the first and most important shipbuilder mentioned in the Bible, at work on the ark. Just as the act of shipbuilding was incidental to portrayals of Noah, so too were any tools and methods that might show up in the work of art. It did not matter if the illustration came from a church door, as with the ivories done for church doors in Salerno in the eleventh century, or from an illustrated Bible, as with Aelfric's Paraphrase done at Canterbury in about 1050, or whether it was a mosaic, as on the wall of the Capella Palatina in Palermo done in the 1150s for Roger II, king of Sicily. In each case it is Noah, his purpose and his connection to events in the Old Testament and the New, that is the object of the artist's interest. The ship is an afterthought. Indeed, most artists who bothered to depict Noah did not bother with the act of building the ship. Though there were differences in the appearance of Noah in northern as opposed to southern Europe and differences in his relationship to the task at hand, the variations are far from central to what the artist was doing. Noah was responsible for building the ship, so artists showed him as a real shipbuilder of his day, reflecting existing practice but doing so unconsciously.[11]

In the High Middle Ages a changing understanding of the place of the natural world emerged which affected the extent and the character of illustration of nature. The discovery of nature became a valued exercise, and its depiction, as a

---

[11] Richard W. Unger, *The Art of Medieval Technology: Images of Noah the Shipbuilder* (New Brunswick and London, 1991).

result, a noble goal.[12] The new status of the study of nature meant also an enhanced status for technology. The greater importance of how to do things and the people who did them appeared most clearly in the treatment of the mechanical arts. They could be perceived, at least from the twelfth century on and most notably by Hugh of St. Victor and other Victorines, as a way of enhancing individual spiritual life.[13] Changes in economic circumstances, rising urbanization and changes in ideas about labor led to increasing depiction of work. Even so, methods of working were still incidental to the principal themes of contemporary artists. It might be true that accuracy, or rather reflecting nature, was now considered valuable and a positive act of devotion, but it did not mean the purpose of art or depiction was to show technology or even work. The early thirteenth-century mosaic of Noah building the ark from the atrium of St. Mark's in Venice, like the painting of the same topic in the fresco in the Church of St. Francis in Assisi, done between 1292 and 1304, was based on earlier works in a tradition which dated back to the first Christian centuries. However, both now conveyed a more authentic collection of workers, each doing his precise job handling a realistic tool.

With shipbuilding, illustration changed in the fifteenth century, as it did with so many other technologies. The first notebooks or sketchbooks in the tradition of Villard de Honnecourt and the book of Konrad Kyeser, sets of illustrations with running commentary, began to appear. The Venetian Arsenal, the most advanced centre for naval shipbuilding in Europe, was the first to produce treatises on how to build ships and the first to produce illustrations, like those of Zorzo da Modon of c. 1440 (fig. 5). His manuscript essay on galley construction is usually attributed to Timbotta. The notebooks from Venice, like the illustrations, were by-products of an effort to find rules for the design of ships. Various measurements and ratios took on new importance and the sketches showed, in an idealized form, where such measurements were to be taken.[14] Artists continued to paint Noah building ships, as with Raphael's ceiling fresco, part of a program of biblical scenes in the Vatican Loggia done in 1518 or 1519, and as with the illustrator of the *Nuremburg Chronicles* of 1493. Such works showed the influence of earlier pictures of Noah in charge of the building of the ark. They also showed the more general influence of thinking about the place of Noah as a creator, in this case of a vessel, and of Noah prefiguring Christ. Noah is an important and even a dominating figure in such depictions of shipbuilding. Artists also seem more taken with shipbuilding methods of the past for their

---

[12] M.-D. Chenu, *Nature, Man and Society in the Twelfth Century: Essays on New Theological Perspectives in the Latin West*, ed. and trans. J. Taylor and L. K. Little (Chicago, 1968), pp. 39–42, 47–48.

[13] George Ovitt, Jr., *The Restoration of Perfection Labor and Technology in Medieval Culture* (New Brunswick, 1987), pp. 116–7.

[14] R. C. Anderson, "Italian Naval Architecture about 1445," *Mariner's Mirror* 11 (1925), 135–163; John Dotson, "Naval Architecture and the Renaissance Art of Design Treatises, Manuscript and Printed, on Shipbuilding up to 1650," in *Cogs, Caravels and Galleons: The Sailing Ship, 1000–1650*, ed. Richard W. Unger (London, 1994).

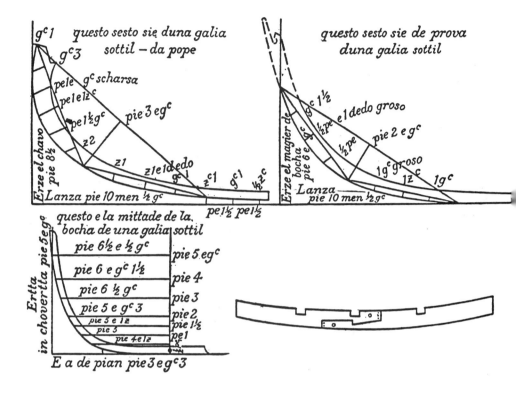

Fig. 5. Diagrams of proportions for the building of a galley from a mid-fifteenth-century notebook done at the Venetian Arsenal. From R. C. Anderson, "Italian Naval Architecture about 1445," *Mariner's Mirror* 11 (1925), 142.

models in creating their images of Noah at work. By the early sixteenth century, though, pictures like those were no longer the only form of illustration of shipbuilding.

In the seventeenth but more so the eighteenth century, protocols for illustration had emerged among shipbuilders. The Dutchman Nicolaes Witsen might in 1671 rely heavily on a Portuguese manuscript of a century before,[15] but he did also turn to contemporaries for information, and he did have precise drawings made under his supervision of imaginary as well as real ships and also of various parts of ships. In all cases he coordinated text and illustrations, making easy reference from one to the other. The Swede Äke Classon Rålamb in 1691

---

[15] Walther Vogel, "Ein neuentdecktes Lehrbuch der Navigation und des Schiffbaues aus der Mitte des 16. Jahrhunderts," *Hansische Geschichtsblätter* 17 (1911), 370–374; Nicolaes Witsen, *Architectura Navalis et Regimen Nauticum ofte Aaloude en Hedendaagsche Scheeps-bouw en Bestier* (Amsterdam, 1690 [1st ed. 1671]).

might pack together in just a very few illustrations a large number of different aspects of the technology of shipbuilding, but the busy plates were the basis for the short treatise which explained in turn what was in the illustrations.[16] By the early eighteenth century a shipbuilder like the Englishman William Sutherland (fig. 6),[17] or in mid-century one like the justifiably renowned Frederik af Chapman, working in Sweden,[18] had in mind a standard method for drawing the lines of any ship. The authors of books on shipbuilding wanted to put the practice on a more mathematical basis. The most successful were French authors, especially Duhamel du Monceau.[19] Such writers not only knew how to make, or have made, accurate, intelligible, and clear drawings, but they also expected, and rightly, that many shipbuilders would be able to understand those drawings. By the 1750s at the latest there was a clear expectation that a drawing of the lines of the ship would and could reflect reality and, even more, that construction could take place based on such an illustration. In the case of the shipbuilder painted in 1633 by Rembrandt,[20] the drawing of a frame for the ship which he is giving to his wife is presumably the guide for construction. That was a first step on the path to the much more precise and more extensive draughts of the eighteenth century. By then the geometric base of "naval architecture," as the topic came to be called by 1700, figured prominently in the illustrations and in their explanations.

The illustration of shipbuilding from St. Peter's Outside the Walls, in that case of Noah building the ark, done in the fifth century (fig. 7),[21] and that in the highly comprehensive polyglot *Allgemeines Wörterbuch der Marine in allen europäischen Seesprachen nebst vollständingen Erklärungen* by the Danish scholar, J. H. Röding from 1793 (fig. 8),[22] show more than that technology changed during the intervening twelve hundred and more years. The illustrations leave no doubt that ideas about the task of building a ship had changed. Drafting, drawing to show what some practitioner is to do, is not the same kind of art. Rather, during those centuries it had become a technology all of its own. Over the years from the later Middle Ages to the start of the Industrial Revolution, the act of shipbuilding came to be divided between design and construction with

---

[16] Äke Rålamb, *Skeps byggerij: eller Adelig öfnings tionde tom* (1691).

[17] William Sutherland, *The Ship-builder's Assistant* (London, 1711).

[18] For example see Dan Harris, "Admiral Frederick af Chapman's Auxiliary Vessels for the Swedish Inshore Fleet," *Mariner's Mirror* 75 (1989), 211–229.

[19] J. R. Duhamel du Monceau, *Elémens de l'architecture navale* (1752); Richard W. Unger, "Design and Construction of European Warships in the Seventeenth and Eighteenth Centuries," in *Les Marines de Guerre Europénnes XVII–XVIIIe siècles*, ed. M. Acerra, J. Merino and J. Meyer (Paris, 1986), pp. 22–23.

[20] A. Bredius, *Rembrandt: The Complete Edition of the Paintings*, rev. ed. H. Gerson (London, 1969), p. 322.

[21] H. Stern, "Les Mosaiques de L'Église de Sainte-Constance à Rome," *Dumbarton Oaks Papers* 12 (1958), 157–218, fig. 19.

[22] Johann Heinrich Röding, *Allgemeines Wörterbuch der Marine in allen europäischen Seesprachen nebst vollständingen Erklärungen* (Hamburg, 1793–1798), *Figuren*.

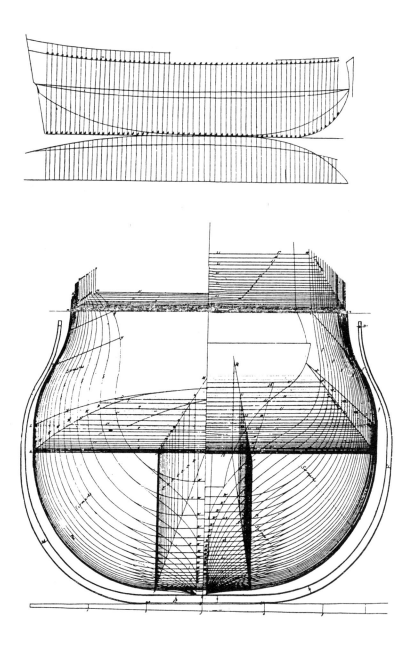

Fig. 6. Sections of a ship showing the form for different ribs at different points along her length. From William Sutherland, *The Ship-builder's Assistant* (London, 1711), in Singer, *A History of Technology*, 3:489.

Fig. 7. Noah building the ark, St. Peter's Outside the Walls, fifth century. From Richard W. Unger, *The Art of Medieval Technology: Images of Noah the Shipbuilder* (New Brunswick and London, 1991), fig. 3.

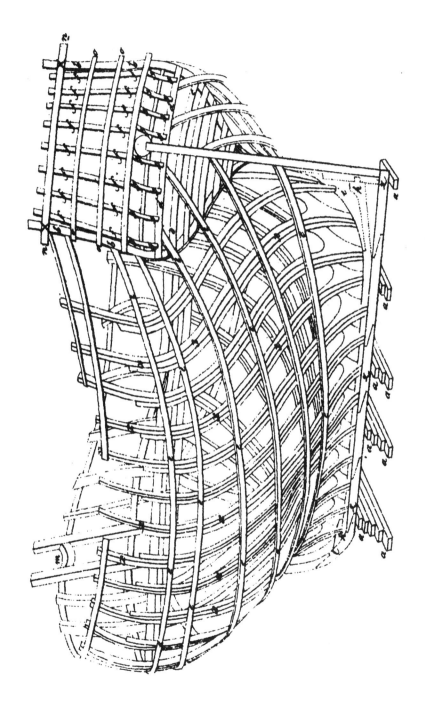

Fig. 8. The frames of a ship from the end of the eighteenth century. From Johann Heinrich Röding, *Allgemeines Wörterbuch der Marine in allen europäischen Seesprachen nebst vollständingen Erklärungen* (Hamburg, 1793–1798), tab. 13, fig. 94.

specialists responsible for each. The tendency toward specialization and the resulting change in the character of work shows up dramatically in the illustrations. Drawings like those in Röding demonstrate the overwhelming confidence in mathematical knowledge to deal with the problem of design, something which turned up already, though tentatively, beginning in illustrations in the fifteenth century.

If illustrations yield more than simply a report of how things are done, then the issue of the accuracy of illustration is equally not a simple matter. Certainly the sketches of sixteenth-century shipwrights give a sense of greater verisimilitude than the art work of about the same time showing Noah building the ark. But that sense may be very misleading. On the one hand, it is easy to forget the long controversy over ship's seals of the Middle Ages and whether their rounded shape did not impose on the artist constraints which forced them to depict overly rounded hulls (fig. 9).[23] Archaeological investigations have shown that, no matter the medium and the shape of the seals, the artists drew ships accurately. On the other hand, the drive for accuracy or consistency of one sort on the part of the artist almost invariably means loss of information of another sort. Draughts like those of Sutherland or Duhamel du Monceau may look clean, neat and orderly, but no ship ever looked like a draught. Not only are the smell and the feel missing from the idealized drawing, but gone too are the mistakes and errors of construction. It is like problems with the recreation of medieval plays where the actors may, after extensive research, get their lines right, but they are healthy, their clothes are clean and they lack the smell of dung. Many features of medieval drama are simply missing. For ships, many features are missing from the draftsman's sketches. Repairs and adjustments were common from the start of the life of a ship, but the naval architects' illustrations do not include any of those additions and corrections. That was true in the eighteenth century even though such draughts by then had become an integral part of the technology of shipbuilding. In the seventeenth century ships had shake-down cruises, and once they were over, ship carpenters usually set to work putting more wood in the right places, often along the waterline to give greater stability, to make the ship seaworthy. Almost immediately the ship itself diverged from the plans. What was drawn as a design to be followed by a technician may never have been acceptable to a captain who had to sail the ship, it may never have worked at all and, in the extreme case, what was drawn may never even have been built. The draught was always more than an ideal but always less than a full, accurate and complete reflection of reality.

---

[23] H. Ewe, *Schiffe auf Siegeln* (Rostock, 1972).

Fig. 9. A ship that appears round on a seal, the oldest seal of Winchelsea, England, thirteenth century. From W. G. Perrin, "Notes on the Development of Bands in the Royal Navy," *Mariner's Mirror* 9 (1923), facing p. 2.

Neither of the naive approaches to illustrations as sources will succeed. Just saying that it is impossible for anything to be reliable or just saying that everything must be accepted at face value[24] makes no progress toward solving the problem of finding out about the history of technology and, for that matter, the history of art. A healthy skepticism is always helpful, but with technology illustrated there are additions to the usual caveats learned by all historians about the treatment of their sources. There is more in the illustration of technology than the thing itself, the object being built or the machine used to make the object. There is as well an idea about the place of the gadget in some larger technology. There are ideas about the technology in some larger social framework. There is, in addition, an idea about the technology itself, its value and nature. Artists report not just the technical history of techniques but, unconsciously, something about the relationship between science and technology and even something about the place of technological activities in the context of other human activities, social and intellectual.[25]

Those pictures of medieval water mills collected with patience and care by Brad Blaine produce information about power sources. In that case as in many others, the illustrations of various technologies collected over the years are of help to scholars endeavoring to understand the ways in which things worked. It may not be until the end of the fifteenth century and the work of Francesco di Giorgio Martini that drawings are precise enough to give unambiguous information on the arrangement of mechanisms.[26] But the earlier pictures can and do produce a lot more than just knowledge about power sources, about how gadgets worked. The depictions of techniques and equipment can be even more helpful to the historian of technology than the clean, neat, tidy and aesthetically pleasing drawings of eighteenth-century draftsmen. The work of the medieval artist may not provide more information, but rather information of a different and equally valuable type for the study of technology, technology in its larger social context. It is a great deal to read from often simple images, but the information is certainly there to be extracted. Such powerful sources deserve a warning label to handle with care.

UNIVERSITY OF BRITISH COLUMBIA

---

[24] Husa, *Homo Faber*, pp. 17–8.

[25] Maurice Daumas, "The History of Technology: Its Aims, Its Limits, Its Methods," trans. A. R. Hall, *The History of Technology* 1 (1976), 89.

[26] Ibid., p. 94.

How Will I Feed My Legions? Provisioning the Roman Army along the Lower Rhine Frontier

by James K. Otte

While rereading Julius Caesar's autobiographical account of *The War in Gaul*,[1] I was struck by his numerous references to grain. If there existed a consistent policy or system of supplying the Roman army in the field, it is not evident from his narrative. Thus I decided to investigate the question Caesar and his successors might well have asked: "How will I feed my legions?"

How did the Romans feed their armies in Gaul, especially along the Lower Rhine frontier? With the greatest concentration of troops along that river, how did the Romans provide the grain for the four legions each in Upper and Lower Germany?[2] That quandary must have been especially vexing along the Lower Rhine frontier, where over a distance of some eighty miles, there were five legionary *castra*, with sixteen auxiliary forts in between them. The standard size of a legion was 5,120 men.[3] In addition, there were some one hundred twenty horsemen attached to each legion who acted as scouts and dispatch riders.[4] Finally, there were the "auxilia," whose strength per unit seems to have been about one-third that of a legion. There are no exact figures for the strength of the auxilia, but since their granaries *(horrea)* are about one-third as large as those of the *castra*,[5] I have taken the granaries' size as the basis for the calculation of their numerical strengths. Put into a simple formula: (5,240/3) (16) = 27,936 men comprising the auxiliaries. The combined total of the four legions (20,960 men) plus the sixteen auxiliaries yields some 48,896 soldiers who had to be fed. Since each legionary received one bushel of wheat per month,[6] almost 50,000 bushels of wheat per month were required to feed the Roman army along the Lower Rhine frontier alone. Could such quantities of grain be grown locally? If not, how did the Romans solve this enormous problem?

---

[1] I used two editions of Caesar's account: *Caesar's Commentaries on the Gallic Wars*, ed. and trans. T. Rice Holmes (London, 1908), and *Caesar: The Conquest of Gaul*, trans. S. A. Hanford, rev. Jane Gardner (London, 1982). All subsequent references to Caeser's *Conquest of Gaul* are from the Gardner revision.

[2] Tacitus, *The Complete Works of Tacitus, Annals* 4, trans. Alfred John Church and William Jackson Brodribb, ed. Moses Hadas (New York, 1942), 5.

[3] Graham Webster, *The Roman Imperial Army of the First and Second Centuries A.D.* (London, 1969), pp. 114–15.

[4] Flavius Josephus, *The Complete Works, The Jewish War* 3, ed. and trans. William Whiston (Chicago, 1899), 6, 2. Cf. Webster, *Roman Army*, p. 116.

[5] Webster, *Roman Army*, p. 216.

[6] Webster, *Roman Army*, p. 32. According to Polybius, "Rations for the legionary were a bushel of wheat per month, for a trooper ten and a half [bushels] of barley and three of wheat. Auxiliaries at this period [c. 200 B.C.] received free rations of the same amounts for infantry, and seven and a half bushels of barley and two bushels of wheat for cavalry."

There has been lively interest in this question, and there has been no paucity of explanations. But many questions remain, as my paper will seek to demonstrate.

CAESAR'S GAUL

In 49 B.C. Caesar took that fateful stride across the Rubicon. In the preceding decade Gallia had become Rome's northern frontier province. Theodor Mommsen formulated its status with these words:

> The boundary of the empire since Caesar's time had been the Rhine from the lake of Constance to its mouth. It was not a [line of] demarcation of peoples, for already of old in the north-east of Gaul the Celts had on various occasions mingled with Germans . . . and Caesar had settled several German tribes along the west bank of the middle Rhine, who . . . indeed adhered more firmly to the Roman rule than the Celtic cantons, and it was not they that opened the gates of Gaul to their countrymen."[7]

By 7 B.C., the Romans had extended their influence to that part of Germany lying between the Rhine and the Elbe—but it had by no means been reduced to tranquillity.[8] Increasingly the Rhine was transformed into a line of defense, and "from the camps on the Rhine the connections . . . ran to the great towns of Gaul and to its ports."[9] East of the Rhine lay "Germania," a region which Tacitus a century later could still describe as a land of forests and swamps, and its inhabitants as keen hunters who lived on plain wild fruit, fresh game and curdled milk. In A.D. 100, the Germans were still a pastoral people who engaged in hunting, while grain formed only a minor part of their diet.[10]

Nor were the vast forests and swamps reported by Tacitus limited to Germany. "We know that northern and north-central Gaul contained a great extent of forest, and that the Ardennes forest extended well beyond its present confines to the north of Luxembourg."[11] The emerging picture of the Lower Rhine frontier is not one of extensive agricultural resources, nor were the pastoral Germans of the land beyond the Rhine prepared to supply the Roman legions with grain. The grain supplies of Narbonensis were too far removed from the army in northern Gaul, and the *Bellum Gallicum* demonstrates that Caesar's roving legions spent much of their time as marauding units in search of food.

Caesar's account of the war in Gaul provides convincing evidence of the difficulties he faced in feeding his legions. In the *Bellum Gallicum* he refers to

---

[7] T. Mommsen, *The Provinces in the Roman Empire* (Chicago, 1968), pp. 25–6.

[8] Ibid.

[9] Ibid.

[10] Tacitus, *The Complete Works of Tacitus, Germania,* trans. Alfred John Church and William Jackson Brodribb, ed. Moses Hadas (New York, 1942), p. 23. All subsequent references to *Germania* are from this edition.

[11] K. D. White, *Roman Farming* (New York, 1970), p. 449.

frumentum, corn or grain, more than eighty times, about once every three pages, and usually in circumstances of dire need. Feeding his legions was without question one of his greatest concerns. He must have quickly learned that an army travels on its stomach. Also, it is apparent that had Caesar's enemies in Gaul or Britain been united, or engaged in a scorched earth policy, his legions could not have survived. His own words provide the testimony. Here are just a few instances.

After Caesar's invasion of Britain in 55 B.C., the Britons quickly recognized Caesar's dilemma. His own account speaks volumes:

> [The Britons] knew that the Romans had neither cavalry nor ships nor grain. . . . They therefore concluded that their best course would be to renew hostilities, cut off our men from corn and other supplies, and protract the campaign till winter, being confident that, if they overpowered them or prevented their return, no invader would ever again come over to Britain. (*Conquest of Gaul* 4, 30)

The Britons wasted no time converting their plan into fact:

> All corn had been cut [by the Britons] except this one spot; and the enemy, anticipating that the Romans would come here, had lain in wait in the woods during the night; then, when the troops had laid aside their weapons and were dispersed and busy reaping, they had suddenly fallen upon them. (*Conquest of Gaul* 4, 32)

Caesar's timely arrival with reinforcements saved the day for the Romans. But in the end Caesar decided to withdraw, blaming his retreat on the approaching equinox, the unsoundness of his ships to risk the voyage later in the season, and by taking advantage of the favorable weather (*Conquest of Gaul* 4, 36).

The situation in Gaul was even more dangerous. There, near the city of Avaricum in 52 B.C., many of the Gallic tribes were united under the leadership of Vercingetorix, who

> . . . watched all our expeditions for forage and corn, attacked our men when they were scattered—for they were obliged to go far afield—and inflicted on them considerable loss, although they took every precaution that ingenuity could devise to baffle him, starting at odd times and in different directions. (*Conquest of Gaul* 7, 16)

Caesar relates how at Avaricum he found himself in a most precarious situation. One of his allies was too poor, he accuses another of negligence, but also admits that the food supplies had been put to the torch:

> Owing to the poverty of the Boii, the slackness of the Aedui, and the burning of the granaries, the army was in the greatest straits of supplies, insomuch that for several days the men were without grain, and only

kept famine at bay by driving in the cattle from distant villages. . . .
(*Conquest of Gaul* 7, 17)

Ever the politician, Caesar uses this occasion to praise his legionaries:

> . . . yet not a word were they heard to utter unworthy of the majesty of
> Roman soldiers with successful campaigns to their credit. (*Conquest of
> Gaul* 7, 17)

But the desperation here and in numerous similar situations is well documented
throughout the *Bellum Gallicum*.

Without question, the problem of supplying the legions along the Lower Rhine
frontier received attention within a generation of Caesar's death. Eight legions
guarded the Rhine frontier. Of these, four were garrisoned along the Lower Rhine
in the five *castra* at Bonna (Bonn), Colonia (Köln), Novaesium (Neuss), Vetera
(Xanten) and Noviomagus (Nijmegen). The three legions of Varus that were
destroyed in A.D 9 by Arminius in the Battle of the Teutoburg Forest and the
legions that in A.D. 43 established Britannia as a Roman province were drawn
from these camps.

The departing legions were soon replaced. As stated above, four legions
(20,960 men) and sixteen auxiliary units (27,896 men), almost 50,000 soldiers,
were the standard military complement that continued to guard the Lower Rhine.
How were they supplied with food?

"MAN SHALL NOT LIVE BY BREAD ALONE"

Many historians of the Roman army have held the position that the ancient
legions were almost exclusively consumers of cereals, which were considered to
have been plentiful. Others have argued that the enormous stores of grain
required to feed the legions could not have been available, especially in the
remote provinces, but also insisted that Roman soldiers only as a last resort
turned to the consumption of meat. The literary evidence seems to support the
second position and, if correct, would point to a supply problem of enormous
proportions. Was there enough grain to supply almost 50,000 Roman soldiers
serving along the Lower Rhine frontier?

The Rhine separated that area of northwestern Gaul from the east or
Germania, whose inhabitants were a pastoral hunting people for whom grain was
only a minor part of their diet. According to Caesar, the Suebi were "by far the
largest and most warlike of the German nations" (*Conquest of Gaul* 4, 1). He
writes: "They do not eat much cereal food, but live chiefly on milk and meat, and
spend much time in hunting" (*Conquest of Gaul* 4, 1). In a celebrated passage
from the *Germania*, Tacitus is even more revealing: "Their food is plain—wild
fruit, fresh game, and curdled milk" (*Germania*, 23). The Germans, according to
Tacitus, found another application for grain. He tells us:

> Their drink is a liquor made from barley or other grain, which they ferment to produce a certain resemblance to wine. . . . They satisfy their hunger without any elaborate cuisine or appetizers. But they do not show the same self-control in slaking their thirst. (*Germania*, 23)

The brew Tacitus described as resembling wine was, of course, nothing other than beer. Even if the Romans of Caesar's time or later centuries had been on amicable terms, it does not seem that the region east of the Rhine could be considered a grain basket. Once again we are reduced to the resources of Gaul, where Caesar's constant search for food hardly leaves the impression that it was a land of plenty.

Then, in a penetrating article published in 1971, Roy W. Davies challenged the established tradition that Roman legionaries had been quasi-vegetarian, that their food had consisted almost exclusively of cereals, and that Roman legions resorted to meat only as a last recourse to avert starvation.[12] Basing his study on literary, documentary and archeological evidence, Davies seems to have destroyed the supposition that the Roman legionary ate only corn and no meat. His thesis holds, in fact, that the Roman legionary's diet was quite balanced. There is much evidence to support that position.

Some of the most convincing evidence arrayed by Davies is based on data compiled from archeological finds at thirty-three military sites in the provinces of Britain and Germany. At these thirty-three locations the bones of ox were found in all, of sheep in thirty, of goat in twenty, of pig and red deer in thirty-one, of roe deer in seventeen, and of boar and hare in fourteen. Bones derived exclusively from wild animals were also present at about half of these sites.[13] For fourteen of these locations and at Maryport, Davies also compiled data for mollusks. These included oysters for all fifteen, as well as edible snails for nine and mussels for eight sites, and a variety of others.[14] A final table comprising eight sites includes data for fowl. Chicken is listed for all eight sites, goose for five, and duck for four. The menu had also featured pheasant, swan, pigeon, raven, dove, eagle, heron, crane and crow and other "exquisite delicacies."[15]

Davies' findings and, therefore, his thesis are not beyond reproach. Here I offer just a few observations.

Of the thirty-one Roman sites in Britain and Germany where animal bones were found, only four were occupied exclusively by legions (Caerleon, Chester, Holt and Vindonissa); three (Rod Hill, Corbridge and Ribchester) had joint tenancy of both legions and auxiliaries, while twenty-four were exclusively occupied by auxiliaries.

We know from the *Bellum Gallicum* that Caesar enlisted both Gallic and Germanic cavalry units, some of which he deployed in the ensuing civil war with Pompey. Since Roman auxiliary units included a veritable collection of ethnicity,

---

[12] Roy W. Davies, "The Roman Military Diet," *Britannia* 2 (1971), 122–42.

[13] Ibid., p. 127.

[14] Ibid., p. 129.

[15] Ibid., p. 130.

the term Roman army diet loses much, if not all of its meaning, when these special forces are involved. Their role is aptly portrayed by Graham Webster:

> All of the non-Roman forces, whatever their status, became known as *auxilia*—aides to the citizen legionaries. . . . The allies of Rome began very early in Republican history to play an effective part in the annual campaigns and large-scale wars. The citizens of Rome provided first-class heavy infantry in the form of legionaries, but in other types of fighting they were not so adept. In particular they did not take so easily to the horse and their own cavalry troops were no match against nomadic peoples nurtured in the saddle. . . . It was not always possible to obtain the required skills from within the circle of accepted allies and so it became necessary to hire mercenaries. Marius used Moorish horsemen in his wars with Jugurtha and Rome recruited archers from the eastern parts of the Mediterranean, slingers from the Balearic Islands and cavalry from Numidia.[16]

Some of the locations from which Davies drew his data were occupied by Roman forces—in the case of Germany ever since the time of Caesar, and in Britain since A.D. 43, when as a result of the Claudian conquest the island was made a Roman province. Some of these camps, so it would seem, must have constituted also the nuclei for Roman garrisons for the next four centuries. During those centuries, however, the ethnic composition of the Roman legions changed rapidly. For instance, at the battle of the Milvian Bridge in 312, the majority of Constantine's army, as well as that of Maxentius, was composed of Germans. There is no reason to think that when a Gaul, a German, a Moor, or a Numidian enlisted in the Roman army that he changed his traditional diet to conform with that of the Roman legionary.

Moreover, it is evident from Gregory of Tours and from Bede that Roman *castra*, once they were deserted by their military occupants, quickly became shelters for the local population. The evidence from the sites compiled by Davies, in other words, may be so "polluted" by non-Roman occupants in the course of four centuries that, indeed, it offers no evidence of the Roman army diet at all. In the absence of a precise date and of the ethnic composition of "Roman" legions in a given locale and at a given time, in the absence of precise dating by Carbon 14, or at least by strata, the compilation of data based on bones of animals and shells of mollusks deserves a healthy degree of the historian's skepticism. These compilations may indeed reflect the ethnic assortment of the Roman legions rather than the diet of the Roman legionary.

Roy Davies died in 1977 at age thirty-five. Had he lived, he may well have revised his Roman army diet. His enormous talent and learning have found a fitting monument in an edition of his collected works.[17] Just how challenging,

---

[16] Webster, *Roman Army*, pp. 142–43.

[17] Roy W. Davies, *Service in the Roman Army,* ed. David Breeze and Valerie Maxfield (New York, 1989).

complicated, and elusive the study of the Roman military diet may remain, has since been confirmed. Without doubt, Davies has broken the spell that once captivated those historians who made the Roman legionary a vegetarian and made him live by bread alone. But the issue is by no means resolved. At Bearsden, Scotland, a Roman camp along the Antonine Wall of the second century, chemical analysis of sewage in the ditch adjacent to the fort's latrine points to the possibility of a primarily vegetarian diet.[18] To that possibility I would like to add the corollary that the vegetarian sewage points to the possibility of a primary Roman garrison, in contrast to an auxiliary.

CONCLUSION

Davies' findings may indeed hold the answer to the question: How will I feed my legions? But did he draw the wrong conclusion by classifying the archeological data as evidence of the Roman army diet? Are those bones and shells the table scraps of Romans?

As we have seen, the Roman army underwent an interminable metamorphosis from a citizen force to a multiethnic military composite. There is no good reason to believe that a Gallic, a German, or any other auxiliary sacrificed his traditional food for a Roman diet based on cereals. Fighting in a unit among his own companions, it is not likely that he followed the dictum, "when in Rome, do as the Romans do," in matters of diet. Frequently living in camps among his own people, he, without doubt, continued to eat and drink what he had consumed as a civilian. The Gaul persisted in his indulgence of pork and the German still savored his "wild fruit, fresh game and curdled milk," washed down with liquefied barley.

The ethnic consideration places the question of the Roman army diet, with its heavy demands on grain, in a new perspective. Therefore, my original formula, 20,960 legionaries plus 27,936 auxiliaries translated into 48,896 lbs., or almost two and one-half tons of wheat per day to feed the soldiers along the Lower Rhine frontier, may not be entirely accurate. Since the auxiliary units were composed largely of Gauls, Germans and other non-Romans, whose diet placed less emphasis on cereals, perhaps only about half of the 48,896 lbs. of grain *per diem* projected above was necessary to feed the legions of Germania inferior. Feeding the Roman legions remained an enormous logistics problem, but the increasing number of non-Romans in the ranks of its army also decreased the burden that had once been so endemic to the supply of grain.

Caesar's legions consisted primarily of Romans, those of Constantine of Germans. When the latter defeated Maxentius at the Milvian Bridge in A.D. 312 their respective armies were composed predominantly of Germans. Without doubt, the army's diet reflected that transformation. The large ever-increasing

---

18 J. H. Dickson, C. A. Dickson and D. J. Breeze, "Flour or Bread in a Roman Military Ditch at Bearsden, Scotland," *Antiquity* 53 (1979), 47–51, plate 10.3. Cf. B. A. Knights, C. A. Dickson, J. H. Dickson and D. J. Breeze, "Evidence Concerning the Roman Military Diet at Bearsden, Scotland, in the Second Century A.D.," *Journal of Archeological Science* 10 (1983), 139–52.

presence of non-Roman auxiliaries—more specifically their non-Roman diet—alleviates, so it seems to me, the logistic burden imposed by a traditional Roman army diet largely based on grain.

UNIVERSITY OF SAN DIEGO

The *Artes mechanicae*, Craftsmanship and the Moral Value of Technology

by Elspeth Whitney

The roots of cultural evaluations of technology are complex, multi-faceted and at least for the pre-modern world, difficult to recover. While it is generally accepted today among historians of technology and others that the Middle Ages "valued" technology in novel ways, both the exact perimeters and the reasons for this development are still much debated. One area of current controversy, for example, is whether medieval attitudes toward the environment can be characterized as "aggressive" or even "exploitative" as Lynn White has effectively argued, or whether it would be more appropriate to describe the dominant attitudes as one of "caretaking and stewardship" as other medieval historians and historians of Western attitudes toward nature have suggested. Should we, again with White, see religion as the primary cause of the medieval West's development of technology, or should we regard this development as taking place for other reasons, with recourse to religious values merely providing a handy justification? Finally, what weight should be given to ideas about technology in the history of medieval thought proper, that is, ideas generated by theologians and philosophers rather than attitudes expressed by more "hands on" types working on manors and monasteries?[1]

In this essay, I would like to focus on a small part of this much larger picture and look at some of the ways in which medieval thinkers attempted to incorporate technology and the mechanical arts into their world view. Such discussion took place largely in the context of discussions of the classification of the arts and sciences. Admittedly, these texts have an abstract and "bookish" quality which tie them to a philosophical tradition rather than to the actual practice of technology. At the same time, I would like to suggest, they provided a contemporary solution to one of the chief problems which has frequently been raised in connection with technological development throughout Western history: how could technology, which so obviously produces instruments of excessive power, wealth and destructiveness be regarded as ethical and moral?

I will focus primarily on the Platonic-Neoplatonic-Augustinian tradition, broadly defined, from the point of view of its relationship to attitudes toward technology and craftsmanship and, in particular, on the somewhat paradoxical suggestion that it was among thinkers in this tradition that the question of the moral standing of technology was most developed. It has often been customary

---

[1] Lynn White, Jr., *Medieval Religion and Technology: Collected Essays* (Berkeley and Los Angeles, 1978); for discussion of White's ideas and related issues see George Ovitt, Jr., *The Restoration of Perfection: Labor and Technology in Medieval Culture* (New Brunswick, N.J., 1987), pp. 12–18, 85–87, 135–136; Elspeth Whitney, *Paradise Restored: The Mechanical Arts from Antiquity through the Thirteenth Century* (Philadelphia, 1990), pp. 1–21, and "Lynn White, Ecotheology, and History," *Environmental Ethics* 15 (1993), 151–169.

and traditional among some historians of technology and science to blame Plato for the radical split between body and mind and corresponding contempt for labor long attributed to pre-modern Western philosophy.[2] Certainly the Platonic claim that the physical world is but a pale reflection of the ideal world of the Forms would not seem conducive to enthusiastic pursuit of the control of nature. Yet the degree to which, prior to the Scientific Revolution, thinkers belonging generally to the Platonic, rather than the Aristotelian, tradition, and even Plato himself, were sympathetic to technology is quite striking. The support provided by a revived Neoplatonism in the Renaissance for the goal of the manipulation of nature, whether by natural magic or technology, is well known. In the twelfth and thirteenth centuries those thinkers who most valued the mechanical arts, including Hugh of St. Victor, Godfrey of St. Victor, Bonaventure, Vincent of Beauvais, and Roger Bacon, also often belonged to a Platonic and Augustinian tradition. Although in the thirteenth century a more Aristotelian tradition on the mechanical arts comes to the fore, this too was influenced by a Neoplatonic strand inherited from Arabic classifications of the sciences. In late antiquity a similar pattern prevailed, with writers in the Neoplatonic tradition most likely to accord at least some crafts a status roughly equal to that of the liberal arts.[3]

The key to the apparent paradox alluded to above lies in two aspects of the Platonic/Neoplatonic tradition: (1) the tendency to link types of human activities in a hierarchy of ascending categories of knowledge, and (2) the tendency to subordinate all knowledge to the over-riding importance of the philosophical and, ultimately, religious goals of human thought and activity. As these two themes were worked out in late antiquity and the Middle Ages, they provided an unusually flexible instrument for including technology within the legitimate and valued (and therefore, moral) categories of knowledge.

The development of a characteristically medieval approach to defining craftsmanship, distinct from manual labor, on the one hand, and, from contemplative science, on the other as I have suggested, took place largely within discussions of the mechanical arts as a part of knowledge. These discussions operated, of course, within a philosophical framework inherited from antiquity. The notion of craftsmanship seems to have posed an exceptionally awkward problem in classical philosophy, in part because it was difficult to fit comfortably within the dualistic opposition of the body and mind which did in fact characterize much of classical thought. Craftsmanship represented to most thinkers an imitation of nature in some sense, and drew upon human capacity for both physical skill and mental ingenuity. Within this general conception, attitudes toward craftsmanship oscillated between a positive "reading" of craftsmanship as a worthy product of human intelligence and a negative "reading" of it as either

---

[2] An extreme example is Benjamin Farrington, *Greek Science: Its Meaning to Us* (London, 1953), pp. 104–111.

[3] See, for example, Allen G. Debus, *Man and Nature in the Renaissance* (Cambridge, Eng., 1978) on Neoplationism and the Scientific Revolution; Henri Irénée Marrou, "Les arts libéraux dans l'antiquité classique," in *Arts libéraux et philosophie au Moyen Age: Actes du Quatrième Congrès International de Philosophie Médiévale* (Montreal, 1969; Paris, 1969), pp. 5–27 on late antiquity.

merely physical or as representing a fraudulent, distracting or deceitful use of the mind and human capacity. On the one hand, craftsmanship was often represented as a kind of unthinking experience (*empeiria*), a mere knack for manual dexterity, substantially different from the rational processes involved in the liberal arts and philosophy. Despite the anthropological interest and appreciation of Democritus in the arts, for example, his comparison of architecture and weaving with the building of nests by birds and webs by spiders comes close to reducing technology to an instinct shared with animals.[4] Similarly, craftsmanship was often regarded as virtually indistinguishable from the simplest manual labor. It was for this reason that Aristotle could justify his descriptions of technological arts as servile and vulgar. Moreover, according to Aristotle, crafts, because they were products of human action, belonged in the realm of the contingent and particular, outside the universal truths of philosophy.[5] On the other hand, most ancient thinkers at the same time assumed crafts to be arts which, like all arts, achieved their purposes by the use of an orderly and rational method. This degree of rationality, however minimal, nevertheless clearly made craftsmanship a part of knowledge in a broad sense and allowed for connections to be built between it and philosophical concerns. Despite Aristotle's often pejorative statements about crafts, for example, his systematic organization of human knowledge, which included the so-called "productive arts," as well as his discussion of the relationship of art and nature allowed a number of Hellenistic and late ancient authors to expand classifications of knowledge to provide a more important place for technological arts.[6] Finally, as Clarence Glacken and others have shown, praise for man's abilities to shape the world through human intelligence and skill and to create, as it were, a "second nature" was a common theme among both Greek and Roman writers.[7]

While the ascription of rationality to crafts linked them, however distantly, with higher types of knowledge, it paradoxically also, however, opened the door to the second charge commonly made against craftsmanship in antiquity and into the Middle Ages: that technological arts were all too successful at allowing and encouraging men to act on their desires for luxury, pleasure and power and therefore led men astray from the ideals of moderation, self-sufficiency and the contemplative life.[8] Sophocles made this point in the *Antigone* when he speaks of the terrible, almost monstrous power of man, who "with the inventiveness of techniques holds in his hand something clever beyond expectation," and is therefore brought to the choice between good or evil. As Arthur O. Lovejoy has

---

[4] Democritus, Fragment 154, in *Die Fragmente der Vorsokratiker*, ed. M. Diehls and W. Kranz, 6th ed. (Berlin, 1954), 2:173.

[5] Aristotle, *Politics* 1337b; *Metaphysics* 981a–b; *Nichomachian Ethics* 1140a.

[6] Whitney, *Paradise Restored*, pp. 36–40.

[7] Clarence J. Glacken, *Traces on the Rhodian Shore: Nature and Culture in Western Thought from Ancient Times to the End of the Eighteenth Century* (Berkeley, Los Angeles, and London, 1967), pp. 3–170; Pamela O. Long, "Invention, Authorship, 'Intellectual Property,' and the Origin of Patents: Notes toward a Conceptual History," *Technology and Culture* 32 (1991), 848–853.

[8] I use the term "men" here self-consciously not as a generic term for humankind but specifically for the male gender.

shown, "nearly all the crafts . . . could more or less plausibly be, and, as the texts show, in antiquity were, brought under an . . . indictment [of promoting greed and violence.]"[9] In contrast to the liberal arts, the banausic arts, a pejorative term often including most, if not all, crafts, were thought to promote sluggishness of mind, tendencies toward luxury and general vulgarity. Keeping in mind that virtue and reason were usually inextricably linked in classical thought, it is not surprising perhaps that activities judged to be barely rational were also judged to be potentially disruptive and probably immoral. Nevertheless, the extent to which ancient writers emphasized the subversive force of crafts to corrupt alerts us to the extent to which crafts were imputed powers not always explicitly acknowledged. Although art imitated nature, crafts, like magic, were suspected of going beyond the natural or, at least, of directing man's attention away from the moral constraints mediated by nature. Yet, at the same time, if the power of craftsmanship could be controlled and directed by the moderating influence of the "natural," then this power might be viewed as both fitting and virtuous.

Plato, of course, was one of the architects of this complex and ambivalent attitude toward technology. For Plato, as for many classical writers, craftsmanship not only was associated with physical, as opposed to mental, labor, but was suspect because it provided means without knowledge of appropriate ends, abilities without the wisdom to govern them.

> Cookery puts on the mask of medicine and pretends to know what foods are best for the body . . . Now I call this sort of thing pandering and I declare that it is dishonorable . . . because it makes pleasure its aim instead of good, and I maintain that it is merely a knack and not an art because it has no rational account to give of the nature of the various things which it offers.[10]

Elsewhere Plato provides us with elaborate comparisons of crafts with the deceptive arts of sophistry and poor rulership.[11] Yet Plato's very insistence that knowledge, to be knowledge, be virtuous at the same time allowed for at least some crafts to be incorporated into the higher aims attributed to the liberal arts and philosophy.[12] Plato occasionally hints in this direction himself. In the *Laws*, for example, Plato makes a passing remark that while the products of the arts of painting, music and other crafts are mere "toys" or "simulacra," there are arts of genuine worth, including medicine, agriculture and gymnastics, which "lend their aid to nature." He then goes on to list statesmanship and legislation.[13] Much later, this basic hierarchy of the arts, beginning with those that produce a sensible

---

9 Arthur O. Lovejoy and George Boas, *Primitivism and Related Ideas in Antiquity* (Baltimore, 1935), p. 7.

10 Plato, *Gorgias* 464d–465a, trans. W. Hamilton (London, 1960), p. 46.

11 Plato, *Sophist* 219b–d and *Statesman* 281d–289e.

12 On Plato's use of craftsmanship as a metaphor for the bringing of order out of disorder, see G. R. E. Lloyd, *Polarity and Analogy: Two Types of Argumentation in Early Greek Thought* (Cambridge, Eng., 1961), pp. 271, 292–294.

13 Plato, *Laws* 889c.

product, passing through those which "lend their aid to nature," and ending with the higher arts of governance, was expanded by Plotinus into a carefully worked out scheme which tied human art to the Intellectual world of the Good. Plotinus begins with the "imitative arts," that is, painting, sculpture, dancing, pantomime and gesturing, which he says are "largely earth-based." Next is music, the earthly representation of the music of the Ideal Realm, followed by crafts such as building and carpentry, which give us matter in wrought forms and draw on pattern, taking their principles from the "thinking There," but bringing these down to the level of the material realm. Still closer to the intelligible world are agriculture, medicine, gymnastics and other arts which "deal helpfully with natural products, seeking to bring them to natural efficiency." Finally, following Plato's order, Plotinus refers to oratory, generalship, administration and sovereignty—which, ideally, look to the Good and are associated with it—and geometry and philosophy, the sciences of true Being. What we have here is a kind of "great chain" of the arts, in which at least some technological arts appear surprisingly high up in the list.[14]

Plotinus also develops a significant distinction between those arts and crafts which merely produce a physical object—that is, the Aristotelian productive arts—and those arts and crafts which more closely mirror the action of the Divine Craftsman because they actively aid the processes of life. All crafts participate to some extent in both the sensible and the intelligible realms. They, in Plotinus's words, "draw upon pattern" but differ according to the closeness of their subjects to the true realm of being. Agriculture, medicine and gymnastics, which deal with living beings, whose life derives from the higher sphere of Life, therefore, are of a higher order than those crafts which have more purely material subjects. This distinction is elaborated elsewhere in the *Enneads*. Plotinus further defines the artificial as follows: "the artificial either remains, as it began, within the limit of the art—attaining finality in the artificial product alone—or is the expression of an art which calls to its aid natural forces and agencies, and so sets up act and experience within the sphere of the natural."[15] Although highly abstract, such a definition of art implicitly makes claims for human technology as powerful as those of any author up to, and possibly including, the Renaissance.

The introduction of Christianity at first made little apparent difference to philosophical attitudes toward technology. When Tertullian responds to Virgil's prophecy in the Fourth Eclogue of a blessed future in which frolicking multi-hued lambs would obviate the need to dye wool by grumpily remarking that if God had wanted us to wear purple clothes he would already have made purple sheep, he is both making a political statement and echoing classical

---

[14] Plotinus, *Enneads* 4.4.31, trans. Stephen MacKenna (London, 1962), p. 314. Other authors of late antiquity similarly created a hierarchy of arts in which some technologically oriented arts, most often agriculture, medicine, architecture, navigation, painting and sculpture, occupied a middle ground between the liberal and banausic or vulgar arts; Maximus Victorinus, for example, characterizes agriculture, gymnastics, medicine, mechanics and carpentry as arts "of the body and soul," see Whitney, *Paradise Restored*, pp. 42–46.

[15] Plotinus, *Enneads* 4.4.31, trans. MacKenna, p. 314.

primitivism.[16] Augustine's often quoted discussion of the arts at the end of the *City of God*, despite its specifically Christian context and ambiance, similarly brilliantly encapsulates much of classical ambivalence toward crafts. Technological arts perfected by the natural genius of man and exhibiting an extraordinary acuteness of intelligence, nevertheless, he says, have purposes which most often seem "superfluous, perilous, and pernicious," providing among his examples ceramics, theatrical spectacles, the trapping of wild animals, drugs, and weapons. Although this passage as a whole has often been read as straightforward praise of the arts, it has always seemed to me to have highly ironic overtones. After a brief but positive review of the liberal arts, for example, it concludes, "last but not least, is the brilliance of talent displayed by both pagan philosophers and Christian heretics in the defense of error and falsehood."[17]

At the same time, of course, Christian notions of progress, providential history and Christian reform in many ways altered conceptions of man's relationship to nature. The physical, natural world, designed by God for human benefit, in a fallen world had been rendered antagonistic to man. Nature itself had deteriorated as a consequence of human sinfulness, and the "reform" of nature might therefore take its place alongside the reform of society as part of Christian work in the world. Together with the Platonic and Neoplatonic notion of a "great chain" of arts, linked in a hierarchy of closeness to the Divine, this view of man's relationship to nature provided an enormously fruitful way of incorporating technology into a philosophical and theological perspective.

Although a systematic and fully developed notion of technology and craftsmanship as a remedy for fallen nature only appeared in the twelfth century, the essential preconditions for this view can be found in Augustine. Suspicious as always of the material world, Augustine nevertheless, like Plotinus, describes craftsmanship as linking human intelligence as much to the divine as to the physical world. He shows clearly the influence of Plotinus in his characterization of certain arts as enabling men to promote the processes of nature. Augustine, for example, several times describes agriculture and medicine as external helps for internal natural forces.[18] In the *De doctrina christiana*, moreover, he explicitly echoes Plotinus. Here he divides corporeal arts into those which manufacture a product, those which result in an action, and those which "display a kind of service to the work of God," namely medicine, agriculture and navigation, thereby quoting Plotinus but substituting God for nature and eliminating the possibly suspect gymnastics.[19]

Augustine's idea of nature as the work of God, however, was qualified by the, to him, historical effect of the Fall. While holding that nature itself is good ("All these—sky, earth, fishes, animals, trees—are not evil; it is evil men who make

---

[16] Tertullian, "On the Apparel of Women" 1.8.

[17] Augustine, *City of God* 22, 24 in *The Fathers of the Church: A New Translation*, trans. Gerald G. Walsh, S.J., Demetrius B. Zema, S.J., Grace Monahan, S.J., and Daniel J. Honan (Washington, D.C., 1954), 24:484.

[18] Augustine, *De genesi ad litteram* 8.9.18, *The Literal Meaning of Genesis,* trans. John Hammon Taylor, Ancient Christian Writers 42 (New York, 1982).

[19] Augustine, *On Christian Doctrine*, trans. D. W. Robertson, Jr. (Indianapolis, 1958), p. 66.

this evil world"), original sin has led to the loss of human control over the natural world. Plants and animals no longer cooperate in God's mandate to Adam to subdue the earth and till it, the roses grow thorns, and even the human body became uncontrollable, not only becoming liable to disease but losing its ability to engage in what Augustine oxymoronically describes as "rational sex." Restoration of the human condition to its proper prelapsarian state, therefore, involves not only a moral and spiritual reformation but also a reordering of the human relationship to nature. [20]

Medieval thought on technology and the mechanical arts as a category of knowledge followed two main lines of development: (1) further amplification of the Augustinian idea of technology as part of the overall human task of restoration, and (2) a consideration of technology as the practical and applied side of theoretical science ultimately derived from Aristotelian ideas by way of Arabic writers on the arts and sciences. Taken together, the significant number of medieval writers who articulated the value of technology as a category of knowledge in one or the other of these two ways illustrates a broad interest in exploring technology's place in human thought. Although much of what we associate with the modern concept of technology, in particular an emphasis on the manipulation and control of the natural world for human benefit justified in purely materialistic terms, is missing from these texts, these twelfth- and thirteenth-century thinkers nevertheless contributed in important ways to the development of Western cultural attitudes toward technology. Not only did the mechanical arts become a standard part of human knowledge, but a coherent and positive rationale for technology as intellectual work was fashioned out of the fragmented and diverse elements inherited from antiquity.[21] Moreover, while medieval thought as a whole retained a certain skepticism about the moral value of an *overemphasis* on technological achievement (as indeed we do today), significant effort seems to have gone into justifying the mechanical arts as morally virtuous and intrinsically good.

Central to the medieval development of the Augustinian view of technology as a remedy for the dislocation of human relations with nature brought about by the Fall is the twelfth-century *Didascalicon* of Hugh of St. Victor.[22] Enormously influential during the High Middle Ages and into the fourteenth century, traces of Hugh's concept of the mechanical arts can be found even in the early modern era. According to Hugh, the Fall, having left us in "a great chaos of forgetfulness," renders men ignorant of wisdom, desirous of evil, and sickened with mortality; the mechanical arts, like all human knowledge, are directed toward making up for

---

[20] Augustine, *The City of God* 2.36. On Augustine and sexuality, see Elaine Pagels, *Adam, Eve, and the Serpent* (New York, 1988), pp. 110–114.

[21] On changing attitudes toward labor and crafts in the later Middle Ages, see Jacques Le Goff, *Time, Work and Culture in the Middle Ages*, trans. Arthur Goldhammer (Chicago, 1980); Ovitt, *Restoration of Perfection*, pp. 137–164; Long, "Invention," and *Le Travail au Moyen Age: Une Approche interdisciplinaire: Actes du Colloque international de Louvain-la-Neuve 21–23 mai 1987*, ed. Jacqueline Hamesse and Collette Muraille-Samaran (Louvain-la-Neuve, 1990).

[22] Hugh of St. Victor, *The 'Didascalicon' of Hugh of St. Victor: A Medieval Guide to the Arts*, trans. Jerome Taylor, Records of Civilization, Sources and Studies 64 (New York, 1961).

this overwhelming disaster. Craftsmanship, for Hugh, is the intelligent observation of nature in order to make a product which will remedy human physical deficiencies. Although Hugh refers to the artisan as "imitating nature," he means this not in the Aristotelian sense of imposing order on matter by a process analogous to that used by nature, but rather that the artisan produces a solution to human inconveniences by imitating similar solutions in nature: He provides the example of the house builder who recognizes the need for peaked roofs by observing the flow of water off the sides of ridged mountains. The work of the artisan, who may work "with nature" or "alone without nature," in one sense parallels the work of the Creator and of nature, on one hand, and also represents in the physical sphere the work of restoration or reform also taking place in the intellectual and spiritual realms.[23]

Hugh's scheme for including the mechanical arts as a part of the religious quest for restoration to our prelapsarian condition proved enormously popular in the succeeding centuries. Virtually every thinker to consider the problem of technology as a part of knowledge in the twelfth and thirteenth centuries was to a greater or lesser extent influenced by it. The core of his ideas was often encapsulated in a popular *topos*, that of the three human evils (ignorance, vice and physical infirmities) and their remedies (learning, ethics and the mechanical arts) which appeared in at least fifteen separate texts in the twelfth century alone and a number of others in the thirteenth century.[24] Many of these authors placed particular emphasis on the morality and ethical value of the mechanical arts. Richard of St. Victor, for example, describes theology, ethics and the mechanical arts as the postlapsarian substitutes for the three original goods bestowed by God: man's creation in the image of God, his similitude to God, and the immortality of his body.[25] Godfrey of St. Victor, using a related metaphor of the practical arts, or ethics, and the mechanical arts as two rivers flowing from Paradise, radically proposes that *all* crafts are in themselves both moral and necessary. Whenever crafts are used for pleasure or immoral activities, this amounts to a misuse, revealing not their true nature but a perversion: "For even though the abuse of armed force is the source of hatred and ill-will, nevertheless the aim of its proper use is nothing but peace and calm. And that which has good as its aim is itself good." Like several other of his contemporaries and near-contemporaries,

---

[23] Ibid., pp. 47–56.

[24] L. M. de Rijk, "Some Notes on the Twelfth-Century Topic of the Three (Four) Human Evils and of Science, Virtue, and Techniques as Their Remedies," *Vivarium* 5 (1967), 8–15; Whitney, *Paradise Restored*, p. 104; another text containing this *topos* has been edited by C. S. F. Burnett, "Innovations in the Classification of the Sciences in the Twelfth Century," in *Knowledge and the Sciences in Medieval Philosophy: Proceedings of the Eighth International Congress of Medieval Philosophy, Helsinki 24–29 August 1987*, ed. Simo Knuuttila, Reijo Tyorinoja and Sten Ebbesen, Publications of Luther-Agricola Society, series B 19 (Helsinki, 1990), 2:25–42; the passage containing the *topos* is on p. 38.

[25] Richard of St. Victor, *Liber exceptionum: Texte critique avec introduction, notes et tables*, ed. Jean Chatillon, Textes philosophiques du moyen âge 5 (Paris, 1958).

Godfrey attributes the original source of human knowledge of the mechanical arts, like knowledge of God's law in general, to Moses.[26]

A less well-known work, formerly attributed to William of Conches but now thought to be by a disciple of his, illustrates the kind of subtle individual re-working of Hugh's ideas which was common in the twelfth century. The text carefully integrates Hugh's ideas on craftsmanship into a discussion of God, the Trinity, demons and the *anima mundi*, reflecting a generally Neoplatonic influence. The author also repeats the notion of the three evils and their remedies with the significant addition of the corresponding professions which practice these remedies. Wisdom, which guards against ignorance, he says, is pursued by men of learning; virtue, which guards against vice, is pursued by men of religion; conveniences, which guard against infirmity, are sought by the businessmen of the world. Elsewhere the author uses the common theme of the body politic to emphasize the importance both of the artisan and the merchant: just as the head is the ruler, the arms represent soldiers; the stomach and knees, artisans and workers; the bones and blood, merchants; and the feet, farmers. The anonymous author is also interesting for his attitude toward magic. While Hugh had taken some trouble to specifically exclude magic from philosophy, magic not infrequently ended up being included in lists of the mechanical arts. The *Philosophia*, for example, considerably softens Hugh's condemnation of magic, and the author, admitting that it is "remote" from philosophy, nevertheless attempts to include it, even associating it indirectly with the liberal arts.[27]

The thirteenth century saw both a broadening and systematization of thought on the mechanical arts. The Augustinian tradition continued to influence many writers on crafts, but it was joined by the newly recovered corpus of Aristotelian and Arabic writings which added a new dimension to philosophical conceptions of crafts as knowledge. The assimilation of Aristotle and his Arabic commentators focused attention on the nature of scientific thought and method; sometimes, if not always, such concerns included discussion of the relationship between speculative science and crafts. Arabic classifications of knowledge, which drew upon a system of classification with its roots in the Hellenistic period combining Aristotelian and Neoplatonic elements, typically reinterpreted Aristotle's division of philosophy into theoretical and practical in ways which emphasized technology. Whereas Aristotle had defined the practical arts to mean those which involved human actions, namely politics, economics and ethics, Arabic classifications organized philosophy as a whole into related theoretical and operative, or practical, arts which manipulated matter. Arts such as mechanics or carpentry, therefore, became in the Arabic scheme instances of the application of theoretical sciences, i.e. practical sciences corresponding to

---

26 Godfrey of St. Victor, *Microcosmus: texte établi et présententé,* ed. Philippe Delhaye, Mémoires et travaux publiés par les professeurs des Facultés Catholiques de Lille 56 and 57 (Lille, 1951; Gembloux, 1951), 56:72–74.

27 (ps.) William of Conches, *Un brano inedito della "Philosophia" di Guglielmo di Conches,* ed. Carmelo Ottaviano (Naples, 1935), pp. 22–34. On the relationship of magic and technology, see William Eamon, "Technology as Magic in the Late Middle Ages and the Renaissance," *Janus* 70 (1983), 171–212.

speculative knowledge. In the hands of Robert Kilwardby and Roger Bacon, this consideration of the method and scope of the sciences, drawn from a combination of Aristotelian and Arabic sources, became a *locus* for explicating the utility of crafts in supplying information and instruments to the theoretical branches of knowledge. At the same time, these sources suggested that crafts be regarded as a basic part of civil life and the maintenance of the community. Arabic classifications of knowledge therefore not only typically included what we would call technological arts as the practical or operative branches of theoretical sciences (for example, listing carpentry and mechanics as practical mathematics), but also suggested that crafts came under the broad rubric of the art of civil life, especially the art of ruling the family. In the West, writers influenced by this tradition might include the mechanical arts, sometimes paired with the *trivium*, or with law, as a division of politics.

In general, thirteenth-century writers on the mechanical arts fall into three clearly distinguishable categories. Some, for example Bonaventure and Vincent of Beauvais, continued the Augustinian-Victorine tradition, making some use of the new Aristotelian terminology but in general adding little of its content.[28] Others, including Aquinas, Albertus Magnus, Michael Scot, and Jean of Antioch, reproduced in truncated forms either Aristotelian or Arabic definitions and classifications of crafts. It is within this group that crafts, following Aristotle, are most often tagged as "servile," and indeed a latent tension can often be detected between a more purely Aristotelian attitude toward crafts as vulgar and an Arabic-inspired more neutral or positive estimation of them.[29] Thomas Aquinas, for example, often emphasized the theoretical, non-utilitarian character of natural philosophy in such a way as to sharply separate crafts from philosophy. Consequently, despite his interest in the autonomous value of the natural world, indeed, perhaps because of it, Aquinas shows relatively little interest in the mechanical arts and, far more than most of his contemporaries, labels them servile and degrading. The final, and most innovative, group, composed of Kilwardby and Bacon, creatively attempted to synthesize the two disparate traditions available to them into a new and comprehensive view of the place of the mechanical arts in human life and knowledge.

Roger Bacon and Kilwardby's approach to the mechanical arts are in many ways the mirror opposite of each other's. Kilwardby retained the form of Hugh's seven mechanical arts but rejected the Victorine function of the arts as means to a spiritual end, preferring the developing secular Aristotelian and Arabic notion of operative crafts as the practical side of the speculative sciences. Bacon, on the other hand, assimilated the Arabic placement of crafts as practical sciences with a

---

[28] Bonaventure, *Saint Bonaventure's De reductione artium ad theologiam: A Commentary with an Introduction and Translation*, ed. and trans. Sister Emma Therese Healey (Saint Bonaventure, N.Y., 1939); Vincent of Beauvais, *Speculum quadruplex sive speculum maius*, vol. 2, *Speculum doctrinale* (Graz, 1965), 1.1–18.

[29] Michael Scot, *Divisio philosophiae* in Vincent of Beavais's *Speculum doctrinale* 1.16; Jean of Antioch, *Notice sur la Rhétorique de Cicéron traduite par Maître Jean d'Antioche, ms. 590 de Musée Condé*, ed. Léopold Delisle (Paris, 1899), p. 14. For Aquinas and Albertus, whose references to the mechanical arts are scattered, see Whitney, *Paradise Restored*, pp. 136–141.

thoroughly Augustinian view of the purpose of knowledge as ultimately moral and religious.

Kilwardby begins his full-scale treatise on the divisions of philosophy by distinguishing among knowledge necessary to salvation, that is, Sacred Scripture, knowledge which is superstitious, injurious and to be avoided, i.e. magic, and science which teaches truths about things or honorable conduct, that is, philosophy, including the mechanical arts.[30] The latter, defined much as Hugh of St. Victor had done, but with improved and streamlined categories, are paired with ethics as the seven practical sciences of the body and soul respectively.[31] At the same time, having quietly abandoned the Augustinian ideal that all knowledge directly serves the purpose of human salvation, Kilwardby turns to a justification of the mechanical arts based on a reworking of Aristotelian and Arabic ideas.

Aristotle, according to Kilwardby, had argued that science by definition must be concerned with universals only, while mechanics, because they proceed from human acts and works, are concerned with singulars or contingencies. Against this radical separation, Kilwardby poses a carefully worked out argument that, despite the distinction between operative and theoretical knowledge, the speculative sciences are practical and the practical sciences speculative:

> Does not, in fact, arithmetic teach how to add numbers to each other and to subtract them from each other, to multiply and divide and draw out their square roots, all of which things are operations? Again does not music teach to play the lute and flute and things of this sort? Again does not geometry teach how to measure every dimension, through which both carpentry and stoneworkers work? Again, does not one show the time for navigation and planting and things of this sort through astronomy? It seems, therefore, that every single science said to be speculative is also practical. It seems, therefore, that the speculative sciences are practical and the practical speculative.[32]

Moreover, the speculative and practical sciences are precisely related through an extension of the Aristotelian principle of subalternation, that is, the speculative sciences provide the underlying principles or causes while the corresponding practical arts provide the empirical examples. Wool-working, arms making, architecture, agriculture and "food science," for example, are "under" and supported by physics, and, similarly, medicine is aided by both physics and astrology and consequently by astronomy.[33] The mechanical arts, therefore, become part of an ascending hierarchy of sciences, linked to both the theoretical sciences and ethics and through them to philosophy. They are, therefore, properly

---

[30] Robert Kilwardby, *De ortu scientiarum* 1.2, ed. Albert G. Judy (Oxford: Clarendon Press, 1976), p. 9.

[31] Ibid., 34–40, pp. 129–133.

[32] Ibid., 41.379, p. 138.

[33] Ibid., 43.401, pp. 139–140.

part of knowledge, and take their value both from the assistance they offer to human efforts to live virtuously and to understand and make use of the physical world.

Bacon, like Robert Kilwardby, conceived of craftsmanship as the practical side of theoretical science and, like Kilwardby, focused clearly on the potential immediate, concrete results of technological activity. While Bacon remains a theoretician, rather than a "hands-on" practitioner, of scientific method, his focus is consistently utilitarian. When he classifies the sciences, for example, he lists the parts of practical geometry as the science of instruments, agriculture, measurements, construction, weaponry, experimental science, medicine, alchemy and astrology. Practical arithmetic includes the use of the abacus and astronomical tables, calendars, weights and measures, the measurement of distances and dimensions, mathematical games, and business, including selling, contracting, leasing, barter, exchange, adjustments, spending and saving. Elsewhere he associates the science of weights, alchemy, agriculture, medicine and experimental science with physics.[34] This arrangement of the sciences clearly recalls Arabic patterns of classification. Moreover, Bacon articulated a vision of the relationship between science and technology in many ways comparable to that of Kilwardby's and which, like Kilwardby, appears strikingly modern in tone:

> All things of such wonderful utility in the state belong chiefly to this [experimental] science. For this science has the same relation to the other sciences as the science of navigation to the carpenter's art and the military art to that of the engineer. For this science teaches how wonderful instruments may be made, and uses them when made, and also considers all secret things owing to the advantages they may possess for the state and for individuals; and it directs other sciences as its handmaids, and therefore the whole power of speculative science is attributed especially to this science.[35]

The intellectual influences on Bacon's utilitarian concept of the aims of natural sciences remain debated.[36] Yet the technological imagination for which Bacon is famous—his visions of burning glasses, incendiary substances, fabulous medicines, flying machines, submarines, cars that go by themselves and other

---

[34] Roger Bacon, *Communium naturalium* 1, in *Opera hactenus inedita Rogeri Baconi*, ed. Robert Steele (Oxford, 1922), fasc 2:5, and *Communia mathematica*, in *Opera*, ed. Steele, fasc. 16:42–49.

[35] Roger Bacon, *The Opus Majus of Roger Bacon,* trans. Robert Bell Burke (New York 1962), 2:633.

[36] N. W. Fisher and Sabetai Unguru, "Experimental Science and Mathematics in Roger Bacon's Thought," *Traditio* 27 (1971), 353–378, attributes much of Bacon's achievement in developing an original scientific method to his attempt to fuse "traditional" Platonism with the new Aristotelianism, 374–375; Steven J. Williams, "Roger Bacon and his Edition of the Pseudo-Aristotelian *Secretum secretorum*," *Speculum* 69 (1994), 68–70, argues against an overemphasis on the *Secretum*, suggesting that the idea that natural knowledge could be translated into practical power was very much "in the air" in Bacon's time.

wonderful instruments—whatever its diverse sources, was ultimately dedicated to the recovery and dissemination of the Christian faith and had its roots in the Augustinian program of human knowledge in the service of salvation. All knowledge, according to Bacon, has been given to man "by one God, to one world, for one purpose," and natural philosophy and experimental science, no less than theology and canon law, are important aids to faith and part of the wisdom of God.[37] Although his notion of the relationship of theoretical and practical science depends heavily upon Arabic thought, Bacon's view of the purpose of knowledge and technology's role must be included within the Augustinian and Victorine traditions.

The thought of Hugh of St. Victor in the twelfth century and Kilwardby and Bacon in the thirteenth represent the most developed medieval thought on craftsmanship and its place in the medieval philosophical and religious world-view. To a large extent, the work of these authors attempted to deal with the questions raised by classical philosophy concerning both the rationality and virtuousness of technological activity which followed from the radical distinction between knowledge, pursued for its own sake and valued for itself as truth, and crafts, by their very nature directed toward a useful corporeal end and subordinate to other kinds of knowledge for their final ends and purposes. The dualism reflected in this separation of mental and manual work tended to split crafts off from the realm of thought and relegate them to the shadowy realm of the "merely" physical. The dichotomy could be overcome, as it was by Robert Kilwardby, by stressing the interdependence of practical and theoretical science. To integrate fully the mechanical arts and natural philosophy, however, stretched the perimeters of contemporary scientific knowledge and, ultimately, required the complete overthrow of Aristotelian science. A simpler and more easily assimilated way of bridging the gap between speculative philosophy and technology was provided by the Platonic-Augustinian-Victorine view of knowledge which conceived of all knowledge as finally subordinate to, or "for," the higher purpose of the soul's progress toward God. Although this conception of knowledge could be used to diminish the value of technology as irrelevant or destructive to spiritual values, many twelfth- and thirteenth-century thinkers used it instead to bring technology into the fold of philosophy by subsuming both philosophy and technology under the same function and, in effect, defining both as utilitarian, that is, as means to an end rather than as ends in themselves. Such an approach brought technology back into relationship with other areas of human thought and legitimized it, even if it did not provide a fully developed rationale which valued crafts in and of themselves.

UNIVERSITY OF NEVADA, LAS VEGAS

---

[37] Bacon, *Opus Majus* 2:36.

From Script to Print . . . and Back

by Joseph M. P. Donatelli

In assessments of the impact of printing upon the manuscript culture of the later Middle Ages, the phrase "from script to print," and the teleological project which it announces, has echoed from various quarters: H. J. Chaytor used it as a title for his book, Rudolf Hirsch as an introductory chapter title, and most recently, Norman Blake has employed a variant, "manuscript to print" as the subtitle of a chapter which fittingly knits up a volume devoted to medieval book culture by calling itself "Aftermath."[1] In an impressive study which is committed to reading the shift to a print culture as both evolutionary and revolutionary, Elizabeth Eisenstein has taken the measure of the ways in which printing effected a wide range of social and intellectual transformations.[2] These "Great Divide" theories, as Ruth Finnegan has termed them,[3] which were very much in vogue during the late seventies and early eighties, not only with reference to relations between manuscript culture and print, but also between orality and literacy,[4] have given way to more nuanced readings of the relation between the media of writing and printing during the fifteenth and sixteenth centuries.[5] This revisionist reading of the complex and reciprocal relations between orality, writing and printing during this period may have come more sharply into view as a result of our own experience with a rich and complex media mix at the end of the twentieth century, an environment which is capable of embracing oral exchanges (both personal and electronic), handwriting, printing, and now, computer-mediated communication, the latter signalling a media cusp which Richard Lanham has spoken to in his article entitled "The Electronic Revolution."[6]

I would like here to re-open the question of the relation between handwriting and printing, with a view to showing the inadequacy of the evolutionary, unilinear, and progressive models of development which are implied by the neat,

---

[1] H. J. Chaytor, *From Script to Print: An Introduction to Medieval Vernacular Literature* (Cambridge, 1950); Rudolf Hirsch, *Printing, Selling and Reading 1450–1550* (Wiesbaden, 1967); Norman Blake, "Manuscript to Print," in *Book Production and Publishing in Britain 1375–1475*, ed. Jeremy Griffiths and Derek Pearsall (Cambridge, 1989).

[2] Elizabeth Eisenstein, *The Printing Press as Agent of Change*, 2 vols. (Cambridge, 1979).

[3] Ruth Finnegan, "Communications and Technology," *Language and Communication* 9 (1989), 114–15.

[4] Perhaps the most important study of this kind is Walter Ong's *Orality and Literacy: Technologies of the Word* (London, 1982).

[5] See Michael Clanchy, "Looking Back from the Invention of Printing," in *Literacy in Historical Perspective*, ed. Daniel P. Resnick (Washington, D.C., 1983), pp. 7–22; D. H. Green, "Orality and Reading: The State of Research in Medieval Studies," *Speculum* 65 (1990), 267–80. See also the essays collected in *Printing the Written Word*, ed. Sandra L. Hindman (Ithaca, 1991), and in *Vox Intexta: Orality and Textuality in the Middle Ages*, ed. A. N. Doane and Carol Braun Pasternack (Madison, 1991).

[6] Richard Lanham, "The Electronic Revolution," *New Literary History* 20 (1989), 265–90.

almost formulaic, phrase "from script to print."[7] Models of one medium supplanting another in historical succession, originally suggested by Marshall McLuhan's work, prove to be incapable of doing justice to the complementarity and accommodation of media in a society.[8] As a technology of the hand, writing continues to exercise its hegemony in certain types of documents and social exchanges. Perhaps the most telling example is the signature, and the notion of personal surrogacy which it inevitably carries.[9] There are also print-objects that invite the act of writing, and thereby establish a relation between the two media in a single, unified textual field defined by the page. Finally, we may consider those documents which according to evolutionary models have been considered retrograde, if not perverse: handwritten copies of printed editions. The relative neglect of these documents, which have too often occasioned "puzzlement" rather than analysis, has been undoubtedly reinforced by recensionist editorial methods preoccupied almost exclusively with the trunk, rather than the farthest limbs, in constructing a *stemma codicum*. With the printed edition in hand, a later manuscript copied from that edition would be duly noted for the sake of completeness but then readily drop out of sight.

Recent studies have advanced the view that previous efforts to construct a binary opposition between handwriting and printing depend upon a characterization of late medieval scribal activity and book production that is unnecessarily and misleadingly reductive. In his essay "Looking Back from the Invention of Printing," Michael Clanchy has emphasized the continuities between the projects of late medieval literate culture and printing: "Instead of viewing printing as the starting point of a new age, I want to look at it as the endpoint or culmination of a millennium. Writing was of extraordinary importance to medieval culture; otherwise printing would not have been invented."[10] During the fourteenth and fifteenth centuries, "textual communities," to use Brian Stock's valuable term, had developed sophisticated strategies for the production, publication, and circulation of manuscript books for an audience of both private and institutional readers. Institutional communities, most notably the university centers as well as the religious houses participating in the reform movements of the fifteenth century, required methods of production, such as the pecia system, which would ensure both the quality and supply of texts for members of those communities.[11] In yet another historical context, the quaint picture of Chaucer

---

[7] For a critique of these interpretations, see Jonathan Goldberg, *Writing Matter: From the Hands of the English Renaissance* (Stanford, 1990), pp. 1–26.

[8] See, for example, the wide-ranging historical observations in Finnegan, "Communications," as well as Deborah Tannen's language-centered discussion in "The Oral/Literate Continuum in Discourse," in *Spoken and Written Language: Exploring Orality and Literacy*, ed. Deborah Tannen (Norwood, N.J., 1982).

[9] See, however, Goldberg's reading of the signature as a social marking of the individual, *Writing Matter*, pp. 233–50.

[10] Clanchy, "Looking Back," p. 8.

[11] See Mary A. Rouse and Richard H. Rouse, "The Book Trade at the University of Paris, ca. 1250–ca. 1350" and "Backgrounds to Print: Aspects of the Manuscript Book in Northern Europe of

berating his scribe Adam[12] has given way to an awareness of the sophisticated and possibly professional production and circulation of English poetry for private patrons by the end of the fourteenth century, though it is still unclear whether we may regard this activity as a commercial enterprise.[13] Evidence of the concern with the "standardization" hitherto reserved for print culture is attested by the production of manuscript copies of John Gower's *Confessio Amantis* (possibly overseen by Gower himself) in a large-format 46-line double column, where there is an impressive consistency in text and layout, with minute agreements extending down to minor details.[14]

In conjunction with this historical revisionism, our reading of writing, both medieval and Renaissance, has been assisted by theoretical studies which have posited that writing is not a neutral technology that can be exchanged for or supplanted by another medium. Rather, any particular script is an instance of what has been called "a writing system," a system which determines communicative strategies, language use, and encodes social relations.[15] Stock's valuable distinction between text and textuality[16] has enabled us to see more clearly how the idea of "text" can project itself into social contexts where there is no writing, a point substantiated by Aron Gurevich's example of the exchange of a piece of parchment which, when bearing a seal, could be used to certify a verbal exchange even though that remained unwritten.[17]

Whereas the graphemic conventions of writing bring with them a certain visual segmentation and analysis of language (spaces between words might be cited as an instance),[18] writing must also be positioned in relation to the voice and the body. It is apparent that much of what was written in the Middle Ages was "meant to be received by the ear,"[19] though it is clear that assumptions about a direct and unequivocal relation between oral performance and text, such as the so-called minstrel manuscripts thought to have been copied from dictation, underestimate the complexity of this relation, misjudging the extent to which

---

the Fifteenth Century," reprinted in *Authentic Witnesses: Approaches to Medieval Texts and Manuscripts* (Notre Dame, 1991), pp. 259–338; 449–66.

[12] "Chaucers Wordes unto Adam, His Owne Scriveyn," in *The Riverside Chaucer*, ed. Larry D. Benson (Boston, 1987), p. 650.

[13] A. S. G. Edwards and Derek Pearsall, "The Manuscripts of Major English Poetic Texts," in Griffiths and Pearsall, *Book Production*, pp. 258–59.

[14] Ibid., p. 260.

[15] For discussion of the elements in a "writing system" and their relation to spoken language, see Florian Coulmas, *The Writing Systems of the World* (Oxford, 1989), pp. 37–54. See also Richard Halpern's *The Poetics of Primitive Accumulation: English Renaissance Culture and the Genealogy of Capital* (Ithaca, 1991), pp. 79–85, for a consideration of the ideological function of literacy and writing during the Renaissance.

[16] Brian Stock, *The Implications of Literacy* (Princeton, 1983), pp. 6–7.

[17] Aron Gurevich, *Medieval Popular Culture: Problems of Belief and Perception*, trans. János M. Bak and Paul A. Hollingsworth (Cambridge, 1988), p. 227, n. 3.

[18] For a discussion of the relation between spoken and written language, see Michael Stubbs, *Language and Literacy: The Sociolinguistics of Reading and Writing* (London, 1980), pp. 116–35.

[19] Green, "Orality and Reading," 277.

writing, reading, and voice were interrelated.[20] In view of the public presentation of texts, Hans Ulrich Gumbrecht has reflected on the importance of the body rather than the book as a site for the construction of meaning during the Middle Ages.[21]

However, rather than seeing the physicality of writing as belonging to this discrete historical period, we may also recognize that the body continues to figure prominently in this technology of communication. As Hans Enzensberger has observed, "Writing is a highly formalized technique which, in purely physiological terms, demands a peculiarly rigid body posture. To this there corresponds the high degree of social specialization that it demands."[22]

This concern with the physicality and social context of handwriting has also informed recent studies of Renaissance copy-books. In these discussions, writing and literacy have been viewed as technologies for social control, a position which media theorists have articulated most fully with reference to twentieth-century mass-media environments. In *Writing Matter*, Jonathan Goldberg seeks to establish the ways in which the training of the hand to write correctly, "right writing" as he puts it, especially as an activity which lays the foundation from which all further learning takes place, may be viewed as a form of indoctrination. The extent to which the description of handwriting is conceived in language laden with social values is not unique to the Renaissance. Johannes Trithemius, in his *De Laude Scriptorum*, a spirited defense of scribal copying by an abbot who was a patron of printing, reiterates the traditional status of writing as *opus Dei*:

> In nulla autem re monachus active perfectioni plus appropinquare sufficit quam si ex caritate divinas scripturas rescribendo invigilet. Nam unde elemosynam dabit, qui nichil habet? Scriptor autem devotus opera misericordie abundancius implere dinoscitur, cuius labor maioris meriti comprobatur.[23]

Trithemius' conception of writing, articulated by means of a Christian rhetoric, situates it as an activity which belongs properly to the social practices of a devotional and meditative life. In yet another social context, Thomas Hoccleve, a fifteenth-century English poet who spent his career as a clerk in the Privy Seal Office, writes about writing:

---

[20] See Andrew Taylor, "The Myth of the Minstrel Manuscript," *Speculum* 66 (1991), 43–73.

[21] "The Body versus the Printing Press: Media in the Early Modern Period, Mentalities in the Reign of Castile, and Another History of Literary Forms," *Poetics* 14 (1985), 215.

[22] "Constituents of a Theory of the Media," in *The Consciousness Industry* (New York, 1974), p. 122.

[23] "Nothing will draw the monk more closely to active perfection than giving himself, for the love of neighbor, to the copying of divine Scripture. The monk cannot give alms since he owns nothing. As you well know, a devout scribe performs abundantly the works of mercy; the results of his labor have even greater value." *In Praise of Scribes, De Laude Scriptorum*, ed. Klaus Arnold, trans. Roland Behrendt, O.S.B. (Lawrence, Kansas, 1974), pp. 56–57.

> A writer moot thre thynges to hym knytte,
> And in tho may be noo disseueraunce.
> Minde, ye, and hande, non may from other flytte
> But in hem moot be ioynt continuance.[24]

Hoccleve, whose work is now the subject of serious study after centuries of neglect, speaks of writing in terms that share much with Renaissance articulations, emphasizing physical orientation, fluidity, and harmony, the latter invoked by the same diction found in contemporary expressions of social accord.[25]

In light of the coherence of writing as a system which is based on a specific segmentation of language and which is held in place by social ideology, one wonders about its relegation to second-class and subsidiary status in descriptions of the "print revolution." To a certain extent, this view of writing has been the result of an endorsement of printing as a civilizing and progressive technology, leading, among other things, to a more universal literacy which has been closely identified with social advance.[26] Yet the "outmoded" technology of writing continues to exercise its influence unchecked in a range of social activities, among them personal correspondence, administrative and legal matters, the keeping of account books, and the composition and circulation of poetry in sixteenth- and seventeenth-century commonplace books. It is valuable to consider the significance of these writing practices in connection with various spheres of social and cultural activity. For instance, a recent study has suggested that the publication and circulation of poetry in manuscript during this period may be read as a deliberate effort to situate this verse closer to the social and performative contexts of occasional poetry, that the use of the hand therefore marks a form of personal appropriation.[27]

I would like now to turn to a number of specific manuscripts and printed texts which attest to the complex and deeply ambivalent relation between writing and printing during the fifteenth and sixteenth centuries. Of course, one of the most prominent relations is the appropriation of medieval bookhands as a model for the gothic type used by early printers, along with a set of practices specific to manuscript production, such as abbreviations, columnar format, and foliation.

---

[24] "A writer must unite three things in himself / And in those there must be no separation. / Mind, eye, and hand may not fly from another, / but there must be a joint countenance." *The Regiment of Princes,* lines 994–98, cited from *Selections from Hoccleve,* ed. M. C. Seymour (Oxford, 1981), p. 35.

[25] For examples of this usage, see entries for the verb *knitten* in the *Middle English Dictionary,* ed. Hans Kurath, Sherman M. Kuhn, et al. (Ann Arbor, 1952–), esp. 5(a).

[26] On the social practices of literacy, and the ideologies in which they are embedded, see Brian V. Street, *Literacy in Theory and Practice* (Cambridge, 1984).

[27] Arthur F. Marotti, "The Transmission of Lyric Poetry and the Institutionalizing of Literature in the English Renaissance," in *Contending Kingdoms,* ed. Marie-Rose Logan and Peter L. Rudnytsky (Detroit, 1991), pp. 21–41. See also J. W. Saunders, "From Manuscript to Print: A Note on the Circulation of Poetic Manuscripts in the Sixteenth Century," *Proceedings of the Leeds Philosophical and Literary Society* 6 (1951), 507–28.

Eisenstein has remarked that the resemblance between manuscripts and printed texts during this period belies a dramatic shift in the mode of production,[28] but the question of the degree to which print co-opts or changes "the writing system" remains, I think, open to further investigation.

In demonstrating the complementarity and accommodation of media during this period, examples which combine writing and print in a single textual field are particularly useful. Printed indulgences may be cited as a case of a print-object which continues to invite writing. These indulgences were meant to be completed by the issuing officer, who inscribed them with the name of the individual, as well as the date and place of issuance. In form, they are not altogether unlike the certificates of guarantee which are filled out after a visit to a muffler shop. An indulgence printed by William Caxton in 1476 (*STC* 14077 c.106) shows the inscription of the names of "Henry Langley and his wife Katherine" in a hand which coincides, almost indistinguishably, with the textura typeface of the printed text, a typeface which, we will recall, had emulated late medieval writing.[29] Attesting to the fluidity of exchange between writing and print, two manuscript copies of Caxton's printed indulgence are also extant.[30]

The practice of annotating texts marginally by hand continues in printed editions, but an interactive format is introduced by the practice of leading lines in schoolbooks, that is, inserting strips of type-metal between the lines of type, so that glosses and translations could be copied interlinearly by the teacher and student. The creation of this "negative space" anticipates that the printed page will serve as a site for writing activity. In discussing the survival of such texts in the University Library at Leipzig, Ursula Altmann observes that the margins and blank spaces have been "densely filled" with writing.[31]

Yet another type of relation between the printed and the scribal is encountered in codices which bind incunabula and handwritten texts together. An example of this phenomenon is found in manuscript 14053/68, preserved in the Bibliothèque Royale in Brussels.[32] In this volume, Bonaventure's *De praeparatione ad missam*, which was printed in Antwerp by Mathias van der Goes before 1486,[33] is followed by *De spiritu Guidonis*, an account of a soul who suffers in purgatory, copied in a cursive hand (fols. 53v–54r, fig. 1).

The difference in media, which immediately strikes the eye, belies a network of possible relations between these two texts: *De spiritu Guidonis* specifically

---

[28] Eisenstein, *Printing Press*, p. 51–52.

[29] A. W. Pollard, "The New Caxton Indulgence," *The Library*, 4th ser., 9 (1929), 89–89. All *STC* references are to A. W. Pollard and G. R. Redgrave, *Short-Title Catalogue*, 2nd ed., rev. W. A. Jackson, F. S. Ferguson and Katharine Panzer (London, 1986).

[30] See K. Povey, "The Caxton Indulgence of 1476," *The Library*, 4th ser., 19 (1938–39), 462–64.

[31] Ursula Altmann, "Bibliography, Books, and Readers," in *Bibliography and the Study of Fifteenth-Century Civilisation*, ed. Lotte Hellinga and John Goldfinch (London, 1987), pp. 78–81.

[32] On folio 189r, the date 1459 is recorded; see *Catalogue des manuscrits de la bibliothèque royale de Belgique* 23, ed. J. van den Gheyn (Brussels, 1903), pp. 305–7.

[33] *Gesamtkatalog der Wiegendrucke*, ed. Kommission für den Gesamtkatalog der Wiegendrucke (Stuttgart, 1968–), 4:4668.

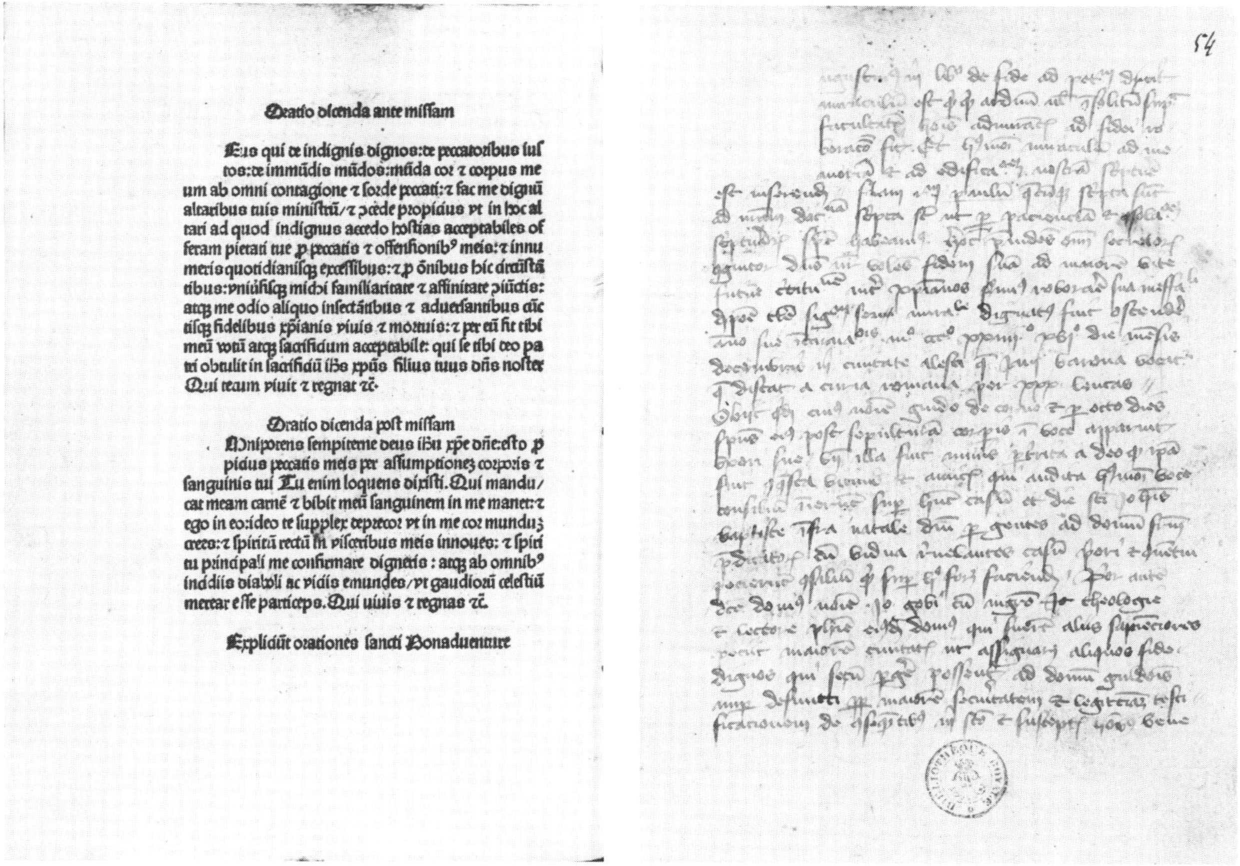

Oratio dicenda ante missam

Deus qui de indignis dignos: de peccatoribus iustos: de immundis mundos: munda cor et corpus meum ab omni contagione et sorde peccati: et fac me dignum altaribus tuis ministrum/ et concede propicius vt in hoc altari ad quod indignus accedo hostias acceptabiles offeram pietati tue pro peccatis et offensionibus meis: et innumeris quotidianisque excessibus: et pro omnibus hic circumstantibus: vniuersisque michi familiaritate et affinitate coniunctis: atque me odio aliquo insectantibus et aduersantibus cunctisque fidelibus christianis viuis et mortuis: et per eum fit tibi meum votum atque sacrificium acceptabile: qui se tibi deo patri obtulit in sacrificium ihesus christus filius tuus dominus noster Qui tecum viuit et regnat etc.

Oratio dicenda post missam

Omnipotens sempiterne deus ihesu christe domine esto propicius peccatis meis per assumptionem corporis et sanguinis tui Tu enim loquens dixisti. Qui manducat meam carnem et bibit meum sanguinem in me manet: et ego in eo: ideo te supplex deprecor vt in me cor mundum crees: et spiritum rectum in visceribus meis innoues: et spiritu principali me confirmare digneris: atque ab omnibus insidiis diaboli ac viciis emundes/ vt gaudiorum celestium merear esse particeps. Qui viuis et regnas etc.

Expliciunt orationes sancti Bonaventure

54

Fig. 1. Brussels, Bibliothèque Royale Albert Ier, MS 14053/68, fols. 53v–54r.
Reproduced by the permission of the Bibliothèque Royale Albert Ier.

mentions the efficacy of prayers which should be recited by the priest before mass,[34] prayers which are prominently displayed as chapter titles in the last leaf of the incunabulum.[35] Also, *De spiritu Guidonis*, copied by hand here, was a text for which there was considerable demand in religious houses, and which was therefore subject to "mass" production, both by scribes and printers, especially during the latter half of the fifteenth century.[36] At this point in the manuscript, the compilation brings into play the resources of written and printed texts, thereby leveling the distinctions which are so apparent when the two texts are considered in isolation from each other, an isolation encouraged by the practice of the segregation of manuscript and printed materials in library archives. The lack of initials, not at all uncommon in utilitarian theological manuscripts such as this one, suggests yet another connection between the texts, showing that both the written and printed text shared equally the inattentions of a rubricator.

After print culture had established itself, the production of copy-books during the sixteenth century affords us another opportunity to reflect upon a complex set of relations between script and print. In his pedagogical manual published in 1612, *Ludus Literarius*, John Brinsley declares that these copy-books might be written out by a scrivener, or, failing that, by the schoolmaster himself, though he makes it clear that the schoolmaster's hand is idiosyncratic and therefore not to be trusted.[37] However, Brinsley considers that the ideals of "perfection" and "constancy" of the hand are best realized by printing:

> If such copie books were finely printed, being grauen by som cunning workman, & those of the most perfect and plaine forms of letters, that could possibly be procured, in a strong and very white paper, one book or two of them would serue a schollar neere all his time, that hee should neuer need to change his hand.[38]

In printed copy-books, there exists a convoluted sequence in which writing is printed as writing for the specific purpose of foregrounding handwriting, and the "perfection" of these printed models was then to be reproduced exactingly by the hands of the pupils. Declaring an ideal which stands apart from the limitations and idiosyncrasies of any individual hand, Brinsley declares that the transfer from print to script was to be so exact that the letters are "as the one hand is to the

---

[34] Bibliothèque royale, MS 14053/68, fol. 60r.

[35] However, the reference in the *De spiritu Guidonis* mentions Ambrose's *Summe Sacerdos,* while the prayer that is appended to Bonaventure's treatise is John Peckham's *Meditatio de Ss. sacramento altaris.*

[36] This work, which survives in more than one hundred manuscripts and three incunabula, circulated widely during the second half of the fifteenth century, when many of the extant texts were copied. For the most complete list of the manuscripts and incunabula, see Thomas Kaepelli, O.P., *Scriptores Ordinis Praedicatorum* (Rome, 1975), 2:2370. Marie-Anne Polo de Beaulieu and I are currently preparing a critical edition of the *De spiritu Guidonis.*

[37] John Brinsley, *Ludus Literarius* (1612; repr. Yorkshire, 1968), p. 32.

[38] Ibid., p. 31.

other."[39] As the analogy draws attention to the body itself, so too do other instructions which tell the writer how to position his body, pen, and eye in relation to his copy:

> For writing straight without lines . . . this may helpe to guide them wel;
> to cause them to hold their elbow so close to their side and so steadily,
> as they can conueniently; for the elbow so stayd, will guide the hand as
> a rule, especially in writing fast.[40]

As a final category, I would like to consider those manuscripts which literally "go back to print" since they have been copied from printed editions. In considering the rationale for such manuscripts, Curt Bühler and Cora Lutz suggested that such copies might be brought into existence by personal commission (especially in light of the elite disdain for the printed book), by the unaffordability or unavailability of printed books, or by personal choice or selection.[41] Bühler notes how exact—his word is "slavish"—such copies could be, and he cites examples of manuscripts which reproduce the identical collation, lineation, and colophons of their printed exemplars. Lutz calls attention to yet another manuscript which is so exacting in its reproduction of a printed exemplar that space has been left for the initials that were never added in the printed text.[42] Yet, on the other hand, a life of St. Winifred in manuscript, apparently copied from Caxton's edition of the *Golden Legend* in the last quarter of the sixteenth century, while declaring in its colophon that it is "drawen out of an ould pryntinge boocke word by word," is in fact a very free rendering of the Caxton version.[43] Blake has pointed out that the large number of manuscript copies of texts excerpted from Caxton's editions may confirm Caxton's good business sense in printing didactic works, such as the *Dicts or Sayings of the Philosophers*, which lent themselves to piecemeal copying.[44] In doing so, Caxton anticipated that his books would serve as a site for scribal interest and activity.

As an example of a manuscript which presents us with a complex set of issues relating to script and print, I would like to cite Douce 261 of the Bodleian Library, a manuscript which collects four metrical romances, all of them apparently transcribed from printed sources.[45] The manuscript is written in a

---

[39] Ibid., p. 32.

[40] Ibid., p. 35.

[41] Curt F. Bühler, *The Fifteenth-Century Book* (Philadelphia, 1960), pp. 36–39; Cora E. Lutz, "Manuscripts Copied from Printed Books," in *Essays on Manuscripts and Rare Books* (Hamden, Conn., 1975), pp. 129–38.

[42] Lutz, "Manuscripts," p. 133.

[43] Curt Bühler, "A New Middle English Life of Saint Winifred," in *Medieval Studies for Lillian Herlands Hornstein*, ed. J. B. Bessinger and R. R. Raymo (New York, 1976), pp. 87–97.

[44] Blake, "Manuscript to Print," pp. 417–18.

[45] For the most complete description of Bodleian MS Douce 261, see Gisela Guddat-Figge, *Catalogue of Manuscripts containing Middle English Romances* (Munich, 1976), pp. 265–66. Maldwyn Mills is completing an edition of the romances in this manuscript for the Middle English Texts series (Heidelberg). British Library MS Egerton 3132A, which preserves a version of the

single italic hand throughout, and the texts are illustrated by pen-drawings which have been brightly colored. It has been suggested that the initials E. B. which appear in the colophon (fol. 25v) allow us to identify Edward Banyster as either the copyist or the patron who commissioned the work, though the connection is tenuous.[46] The year 1564 is recorded in yet another colophon at the end of the manuscript (fol. 48v), providing a *terminus ad quem* for the copying. The manuscript preserves the sole extant text of the *Jeast of Syr Gawayne*, as well as versions of *Syr Isenbras*, *Syr Degore*, and *Syr Eglamore*. The repeated printing and copying of the latter three texts during the first half of the sixteenth century—*Sir Eglamour* was also to achieve popularity as a broadside ballad—attest to a continuing audience for these works, and the texts of Douce 261 are very close to the printed editions. Though the Douce romances demonstrate remarkably close textual agreement with the extant printed editions, there are a sufficient number of textual variants to raise doubts about whether the Douce texts have been copied directly from any one of these.[47] The Douce *Syr Eglamore*, for example, is apparently most closely related to Richard Bankes' edition, printed c. 1528 (*STC* 7542.5), though certain lines also suggest a connection with the oldest printed edition by Wynkyn de Worde dated c. 1500 (*STC* 7541), though the survival of both of these prints in a fragmentary state qualifies these observations.[48] *Syr Degore* demonstrates repeated agreements with the edition published by William Copland (*STC* 6472.5), who seems to have published a series of metrical romances during the mid-sixteenth century, but the Douce text again apparently does not stand in direct relation to the printed edition.

Yet although it is difficult to establish an unequivocal exemplar, the layout and writing of Douce 261 attest to the influence of the printed book. The copyist has sought to replicate the model of the printed page, while at the same time translating it so that he will be able to reproduce certain features of print with his hand. He renders the black-letter of his exemplars into a very neatly formed italic hand, and achieves, with his careful alignment of thickened capitals at the beginning of lines, an effect which is not unlike the "standardization" of upper-case black-letter type at the left margin. The treatment of decorated capitals, which are used to mark sections of the text, also mimics the treatment of such capitals in print. As in the printed editions, a capital *G*, drawn on a square field which has been carefully ruled, marks the beginning of the account of Sir Degare's unwitting marriage to his mother, even though, as we shall see, this section is fragmentary in Douce (fol. 9r, fig. 2; cf. de Worde's edition of *Syr Degore*, fols. 32v–33r, fig. 4).

---

metrical romance *Robert the Devyll,* is copied in the same hand. See R. Flower, "The Manuscript of the Poem 'Roberte the Deuyll,'" *British Museum Quarterly* 9 (1934–35), 36–38.

[46] M. C. Seymour, "MSS. Douce 261 and Egerton 3132A and Edward Banyster," *Bodleian Library Record* 10 (1978–82), 162–65.

[47] For the textual relations of the extant manuscripts and prints of *Syr Eglamour*, see Frances E. Richardson's edition, Early English Text Society, o.s., 256 (London, 1965), and Gustav Schleich's edition in *Palaestra* 53. For *Syr Degore*, see Gustav Schleich, *Sir Degarre* (Heidelberg, 1929).

[48] Richardson, *Early English Text,* pp. xii–xiv, xix–xx.

Than was there muche noyse and crye
The kinge was soore afshamed for thy
Well I wotte hys doughter was forye
For than fhe wyst yt redvlye
That fhe fhoulde maryed be
To a man of ftraunge countre
And leade her lyfe with fuche a one
That fhe wyst neuer from whence he come
The kinge fayed to fyr Degoer
Come hyther fayre fonne me before
And thou were as gentyll a man
As thou femest to loke vpon
And then coulde wytt and reafon do
As thow arte doughtye man to
I woulde thynke my lande well fett
Yf yt were fyue tymes bett
For worde fpoken I mufte nedes holde
Before my Barons that be fo bolde
I take the my doughter by the hande
And feafe the in all my lande
To be myne heyre after me
In ioye and blyße for to be.

Reate ordynaunce was there wrought
To the churche doore they were brought
And there wedded were verament
Vnto the holy Sachramente
Loke what folye happened there
That he fhoulde wedde hys owne mother
The whiche had borne hym on her fyde
And yet he knewe nothynge that tyde
He knewe nothynge of her kynne

Nor she knewe nothynge of hym
And bothe togethet ordeyned to bedde
Yet parauenture they maye be sybbe
Thus dyd Syr Degore the bolde
He wedded hys mother to haue and to holde
But yet he lett them not somme in feere

Fig. 3. Oxford, Bodleian Library, MS Douce 261, fol. 9v.
Reproduced by the permission of the Bodleian Library.

And thou warre as gentyll a man
As thou semest to loke vpan
And thou coude wyt and reason doo
As thou arte doughty man to
I wolde thynke my londe well beset
And yf it were true tymes the bet
For twoe spoken I must nedes holde
Before my barons that be so bolde
I take the my doughter by the honde
And seese the in all my londe
To be my heyre after me
In Ioye and blysse for to be

¶ How syr Degore wedded his moder the kynges
doughter of Englonde and howe she we knewe ꝑ
he was her sone by the gloues. ꝭꝭꝭ ✶

¶ Here ordynaunce was there wrought
To the chyrche dore they were brought
And were there wedded veramente
Vnto the holy sacramente
Loke what foly happened there
That he sholde wedde his owne modere
She whiche had borne hym one her syde
And yet he knewe nothynge that tyde
Ne knewe nothynge of her kenne
For the knewe nothynge of hym
And bothe togyder ordeyned to bedde
Yet paraventure they may be spredde
Thus dyde syr Degore the bolde
He wedded his moder to haue and to holde
God suffred moche thynge there
But yet he lete them nor synne in fere,
It passed on the lye tyme of none
And the daye was nere hande done
To bed was brought bothe the and she
With grete myrthe and solempnyte
Syr Degore stode and be helde than
And thought of the hermyte holy man
That he sholde neuer for thy
Wedde no wedowe nor lady
But yf she myght the gloues two
Lyghtly vpon her handes doo
Alas than sayd syr Degore
The tyme that euer I was bore
And sayd anone with heuy chere
He had leuer than all this kyngdome here
That is now feffed in to my honde
That I were fayr out of this londe

Fig. 4. New York, Pierpont Morgan Library, PML 21135, fols. 32v and 33r.
Reproduced by the permission of the Pierpont Morgan Library.

The standardization of the lines also corroborates a printed exemplar, for a full page in the manuscript counts thirty-two lines consistently, thereby agreeing closely with the lineation (e.g., thirty-two lines in de Worde and Copland) in the printed editions.

The illustrative program of Douce 261 is almost certainly based on woodcuts or engravings which originally appeared in printed editions. Although woodcuts are of limited value in establishing direct transmission, we may note a particularly interesting correspondence between the illustration of Sir Degare's wedding in Douce 261 (fol. 9v, fig. 3) and the woodcut which depicts the same scene in de Worde's edition (sig. fol. 32v, fig. 4).

If de Worde's woodcut did serve as the model for Douce, it is interesting to observe that the figures too have been "translated" since they wear ruffs as collars in a style contemporary with the mid-sixteenth-century date of the scribe's activity.

All the texts in Douce 261 suffer from extensive lacunae, and even this negative evidence perhaps registers the influence of printed models. The loss of the first thirty-two lines which open *Syr Eglamore*, for instance, may be attributed to a leaf that was torn out of the exemplar, possibly for the woodcut that decorated the title page on the recto side. Other omissions also suggest lacunae in the copyist's printed exemplar. In the folio from *Syr Degore* which contains the illustration of the wedding scene, the Douce version alone among extant versions misses the episode relating the discovery of Degore's marriage to his mother. Without the benefit of an exemplar it is difficult to assess the number of lines which Douce drops after completing a couplet with a line that does not appear in any other printed edition or manuscript: "But yet he lett hem not synne in fere / Hyt drewe towarde the tyme of dynere" (Copland's edition, for example, reads "But yet he lett hem not synne ifere / It passyd on the hye tyme of non").[49] Yet, reading against the Copland edition, Douce's omission of one hundred and twenty-eight lines may perhaps warrant the assumption that two leaves containing thirty-two lines per side were missing in the printed exemplar.[50] In thinking about the scribe's, reader's, or auditor's experience with this anthology, one might reflect on how the writing, with its stylization and standardization, effaces these textual gaps, though obviously these large lacunae would have seriously hampered the sense of the text. The manuscript behaves as if the texts were commensurate to the regularity and clarity of the script. We might also observe that the Douce drawing of the wedding scene offers a surrogate "text" which is missing in the verse because of the lacunae in the verse.

As an anthology, Douce 261 represents the personal appropriation of originally separate printed texts in a single field of textuality, the unity of which is achieved here by the italic hand—a hand in vogue among the upper classes during this period[51]—as well as by the illustrative program. This field, created by

---

[49] Schleich, *Sir Degarre*, p. 97, lines 628–31.

[50] Ibid., pp. 97–105, lines 632–773.

[51] Alfred Fairbank and Berthold Wolpe, *Renaissance Handwriting: An Anthology of Italic Scripts* (London, 1960), pp. 28–34.

hand, "knits"—to use Hoccleve's term—different verse forms, for the romances of Douce 261 are composed in both tail-rime and couplets.[52] The scribe's emulation of printed editions also, I would suggest, brings the "public" nature of the print-object into the realm of the personal hand, for the manuscript gives every indication of having been carefully conceived as a specific and coherent project.

Douce 261, and the printed editions from which it was transcribed, also provides an opportunity to reflect on what Jerome McGann has termed the "textual condition"[53] of medieval texts in a sixteenth-century context. These medieval romances have been recontextualized, and in many ways recomposed, in the printed editions of the sixteenth century. As Marshall McLuhan once observed, the sixteenth and seventeenth century saw more of the Middle Ages than the Middle Ages ever did.[54] Douce 261 as well as other manuscripts such as the famous seventeenth-century Percy Folio (British Library MS Add. 27879), represent yet another recontextualization, one which looks back not to medieval texts themselves, but rather to the textual horizons and surfaces which had been established and defined by sixteenth-century printers rather than by medieval authors and scribes.[55] Handwriting remains, as it did in the Middle Ages, a viable technology for such transfers. In Douce 261, a scribal hand is capable of rendering medieval texts, as they are preserved in black-letter, into a bookish and elegant italic which belongs to the "writing system" of the sixteenth century and yet is but one historical moment in a continuous and highly socialized technology of writing.

"From script to print" is neither a Darwinian narrative nor a myth of Jupiter supplanting Saturn. The desire to see our own heritage in the culture of print, perhaps because of our heavy investment in this medium, has caused us to mitigate the complex and enduring relationship between writing and printing. In her magnificent work, Eisenstein devotes but a few pages to the subject. Our understanding of the media situation of any given historical period must recognize the wide range of socially inscribed sets of relations which exist at any given time between writing, printing, and the spoken word. It would appear that the multi-media situation of our own time has much to teach us about technologies of communication during the fifteenth and sixteenth centuries.

UNIVERSITY OF MANITOBA

---

[52] As noted above, this field also extends to the other manuscript copied in this hand, Egerton 3132A.

[53] Jerome McGann, *The Textual Condition* (Princeton, 1991), pp. 3, 8–10.

[54] *The Gutenberg Galaxy* (London, 1962), p. 142.

[55] On the medieval texts copied from printed editions in the Percy Folio manuscript, see Gillian Rogers, "The Percy Folio Manuscript Revisited," in *Romance in Medieval England*, ed. Maldwyn Mills, Jennifer Fellows, and Carol M. Meale (Suffolk, 1991), pp. 39–65; and Joseph Donatelli, "The Percy Folio Manuscript: A Seventeenth-Century Context for Medieval Poetry," in *English Manuscript Studies 1100–1700*, ed. Peter Beal and Jeremy Griffiths (London, 1993), 4:114–33.

Cantare per Litteras: Hucbald's Design for Chant Notation

by Richard J. Wingell

One of the most important achievements in the field of music during the Middle Ages was the gradual development of a sophisticated system of musical notation, capable of recording and transmitting large repertories of music. Our system of notation is one of the fundamental aspects of European art music, created *ex nihilo* during the Middle Ages, part of the legacy left by medieval artists and thinkers that later ages take for granted.[1] Even if we assume, as now seems likely, that many concepts basic to the European musical tradition, such as polyphony, instrumental music, and settings of vernacular poetry, were borrowed from Islamic culture, our notation system is a European invention. Although there are other cultures that have notation systems, the concept of a system that can convey nearly every aspect of a musical idea, so that one can more or less recreate the musical experience at sight, is a European concept, and the tradition of European art music would never have developed along the lines it did without this fundamental medieval achievement. The idea of design and production in the Middle Ages immediately brings to mind a picture of a medieval *scriptorium,* where scribes and notators selected for their special skills spent every work period, month after month, year after year, laboriously copying words and music. The most important work these monks did was, of course, to chant the Mass and Office—the *Opus Dei,* in Benedict's term—but the production of books, including musical manuscripts—graduals, antiphonaries, tropers, and tonaries— was one of the most significant accomplishments of the monastic system.

Since a sophisticated system of musical notation is so fundamental to the evolution of European art music, it is hard for modern musicians to realize that the process of creating accurate notation was neither obvious nor easy. In this paper I want to focus on the efforts of one man to design a system of notation for chant that would meet the changing needs of his day. The man is Hucbald of St. Amand; his efforts to design pitch-specific notation produced an ingenious if not altogether successful solution to the problem of writing music down in an efficient and clear way. Understanding how one musician grappled with this problem in the early tenth century will shed light on the long and difficult process of designing a suitable and effective system of musical notation.

To understand Hucbald's contribution, we first must understand the historical context in which he worked and the circumstances that led him to confront this problem. Hucbald was a well-known musician, a monk of the monastery of St. Amand at Gembloux, in Belgium.[2] He was active at a crucial time in the history

---

[1] For a general history of medieval notation, see David Hiley, "Notation," Part III, in *The New Grove Dictionary of Music and Musicians,* ed. Stanley Sadie, 20 vols. (London, 1980), 11:344–354; and Solange Corbin, "Neumatic Notations," II–IV, in *New Grove* 11:128–144.

[2] See Rembert Weakland, "Hucbald as Musician and Theorist," *Musical Quarterly* 42 (1956), 66–84.

of music, when the *locus* of musical training was shifting from the monastery to the cathedral school. The establishment of cathedral schools, where systematic education could be provided for the secular clergy, was one of the accomplishments of the Carolingian Renaissance; this development also produced a radical change in the teaching of music. These schools suddenly faced the problem of teaching the chant repertory to students in a setting much different from the monastery, where the system of learning chant by apprenticeship and oral tradition had worked perfectly well for centuries. The cathedral schools did not have long spans of time to devote to learning the traditional chant repertory. To meet this new problem, bishops naturally turned to the best monastic musicians for help in establishing music programs at their schools. Hucbald and his colleague Remi of Auxerre were recruited by Fulco, Bishop of Rheims, to establish the music program in his diocesan cathedral school. In response to this request, Hucbald wrote a famous text on music, usually called *De institutione harmonica,* although the heading in the best surviving manuscript copy is simply "*Incipit Musica Hubaldi.*"[3] Like most writings on music from this period, Hucbald's work is not a speculative treatise, but a book of instruction, a textbook for a music course in a cathedral school.

A century later, in the early eleventh century, Guido of Arezzo, after being ejected from his monastery, was hired by Bishop Theobald of Arezzo to write a course in music for his school; the result was the famous *Micrologus,* the best known and most widely circulated music textbook of the period.[4] By Guido's time, notation had progressed to the point that he could boast in his prologue that he could teach anyone to sight-read the entire chant repertory correctly in a month.[5] He showed little respect for his former colleagues, asserting that the old monks, even after singing by ear for fifty or sixty years, still did not understand what they were doing, and, if they happened to sing the chant correctly, they did so by sheer luck, not by actually knowing anything about music.

Hucbald confronted the problem of notating chant a century earlier than Guido, when this problem of teaching chant in a new way was just beginning to surface. Hucbald's effort was the difficult first step in a shift from oral to written tradition, a change forced on musicians by the circumstances of cultural history. Hucbald was a skilled musician, even named in contemporary records as a composer of specific chants, a rare honor in the monastic world. The notation he knew worked quite well for monks who already knew the chant melodies, but it was not adequate for this new situation, in which a different kind of student had to learn chant quickly.

---

[3] The treatise is preserved in MS Brussels, Bibliothèque Royale de Belgique, 10078/95; an edition appears in Gerbert, *Scriptores ecclesiastici de musica* 1:125–147. A translation appears in Warren Babb, *Hucbald, Guido, and John on Music,* Music Theory Translation Series 3 (New Haven, 1978).

[4] For Guido's life, see Josef Smits van Waesberghe, *De musico-paedagogico et theoretico Guidone Aretino* (Florence, 1953). For a critical edition of the *Micrologus,* see Smits van Waesberghe, ed., *Guidonis Aretini Micrologus,* Corpus scriptorum de musica 4 (1955); for a translation, see Babb, *Hucbald, Guido, and John.*

[5] See *Micrologus,* CSM 4, p. 85.

To understand the notation problem as he saw it, we need to take a close look at the notation of Hucbald's time. Below are three examples of early chant notation. For the sake of comparison, they all contain the same chant, the Gradual *Haec dies* with its verse *Confitemini Domino,* from the Proper of Easter Sunday. Musicians will immediately recognize this famous chant; the polyphonic settings of this piece by Leoninus and Perotinus are well known, and in the middle of the verse is the melisma on the words *"in seculum,"* which became a popular tenor for motets in the thirteenth century.

Figure 1 is a folio from St. Gall 359, a *cantatorium* or soloist's chant book from the monastery of St. Gall in Switzerland, dating from about 850, the oldest surviving notation that we know of.[6] Kenneth Levy argues that there must have been earlier manuscripts containing notation; he reasons that Charlemagne would have ordered the preparation of notated exemplars of the official version of the chant, companion manuscripts to the text exemplars of his time that have survived.[7] If there were such notated manuscripts emanating from Charlemagne's court, the surviving examples of early notation would all have been copied from these official written models. Levy's argument is based on conjecture, and not everyone agrees with his theory; in any case, whether or not earlier notated manuscripts once existed, figure 1 represents the oldest surviving European notation.

Before we look in detail at this example, it is useful to turn for a moment to the work of Leo Treitler, who has published several studies on early notation.[8] According to Treitler, it is difficult for modern readers to appreciate this notation, since we tend to view it in much the same way as we study later manuscripts, expecting the notation to convey the same information that later notation does. Treitler also suggested, in an article, which he published in 1982, that the theory of semiotics provides a useful framework for understanding notation, especially early notation.[9] He bases his discussion on the theory of signs as explained by the American philosopher Charles Peirce; his discussion starts with Peirce's classification of three ways in which signs function—the symbolic, iconic, and indexical modes. For those not familiar with these terms, signs function in symbolic mode when their connection with the signified reality is arbitrary and conventional, so that one must be taught the code before he can read the signs correctly. Signs function in iconic mode when there is some similarity or isomorphism between the sign and the thing signified, so that even an untrained reader can grasp some connection. Signs function in indexical mode when their purpose is to cause an immediate physical response. The three modes can be illustrated by the example of a "men at work" sign a driver might encounter on a

---

[6] *Paléographie musicale,* 2nd ser., vol. 2, plate 87.

[7] Kenneth Levy, "Charlemagne's Archetype of Gregorian Chant," *Journal of the American Musicological Society* 40 (1987), 1–30.

[8] A list of Treitler's publications on this question, a summary of his views, and critical reaction may be found in Peter Jeffery, *Re-envisioning Past Musical Cultures: Ethnomusicology in the Study of Gregorian Chant* (Chicago, 1992), pp. 11–50.

[9] Leo Treitler, "The Early History of Music Writing in the West," *Journal of the American Musicological Society* 35 (1982), 237–79.

mountain road. The words "men at work" function in symbolic mode—the letter shapes are meaningless, unless the driver has been trained to read English.

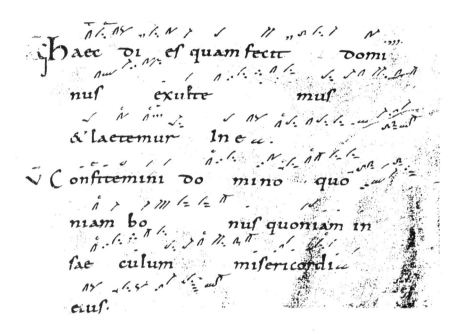

Figure 1

Nowadays, the sign would probably include, in addition to the words, a silhouette depicting men working on a road; the picture functions in iconic mode, since it represents the event pictorially, and communicates something to all drivers, without requiring knowledge of any linguistic system. Finally, the sign functions in indexical mode, since all drivers know that the purpose of the sign is not merely to convey information; it also demands an immediate reduction of speed, since the driver probably will have to come to a stop around one of the next blind curves. As Treitler explains, the three representational modes are hierarchically arranged in the functioning of every system of signs—musical signs as well as other signs—and "the exact hierarchical arrangement will depend on the way the system is used, the nature of the referent field, and the competencies of the readers."[10]

---

[10] Ibid., p. 241.

Applying this theory to the music notation in figure 1, we see that this kind of notation functions in the symbolic mode. Notice the first two signs over the word "*Haec*": first, there is a rounded stroke like an upside-down U, followed by a slash with two dots under it. A reader has no way to interpret those signs unless he has been trained to read the code. A trained reader knows that the initial curved stroke represents a *clivis,* a two-note descending neume, and the slash-and-dots figure signifies a *climacus,* a three-note descending neume.[11] Beyond the arbitrary code for the individual signs, note that there is no "isomorphism" between notation and the course of the melody; that is, the notation lacks iconic function, at least regarding the melodic curve. If we accept the theory that this notation is cheironomic, we might posit a similarity between the shapes of the notation and the gestures of the *precentor* as he leads the singing, but the signs bear no similarity to the shape of the melody.

Note that this notation has no way of denoting pitch, relative or absolute. We know that the melody starts with a *clivis,* but we do not know whether the descending interval is a second, a third, or a fourth; as we move to the second neume, the *climacus,* we have no way of knowing whether it starts lower than the *clivis,* higher, or at the same level. To the surprise of modern musicians, at their first encounter, this notation conveys only one dimension of musical information—the proper series of neumes over each syllable, or the relationship between text and music. Treitler raises an interesting possibility that might not occur to us, since our modern notions of what notation ought to convey are so different from medieval expectations. Perhaps an experienced singer, knowing what Treitler calls the "syntax" of each mode, the *formulae* that are common to each mode, and the way each mode moves around its range and builds coherent melodies, would find in this notation sufficient information to re-create the traditional chant melody, or a melody very similar to it.[12] Whether or not this is true, the problem facing Hucbald remains—this notation is not adequate to convey a melody accurately to the readers he is concerned about, the new students of chant in a cathedral school.

There is another question this notation raises. One might assume, looking at this page, that it tells us something about how the medieval scribe viewed the relative importance of text and music. Since the text seems to be copied more carefully than the music, we might conclude that the words were viewed as the sacred text, to be copied faithfully, and that the musical notation is an afterthought. But that view cannot represent what the scribes thought, despite the crowded and somewhat haphazard appearance of the musical notation. Note that the scribe who copied the text knew very well where the melismas occur in the text and left space for them—in other words, the purpose in the text scribe's mind was clearly to write the text in such a way that the music could be added later. He

---

[11] The most complete tables of neume formations are those found in Dom Gregory Suñol, *Introducció a la paleografia musical gregoriana* (Montserrat, 1925; French translation, Tournai, 1935); cf. p. 187 for St. Gall notation, p. 154 for Laon notation, and p. 163 for Aquitanian notation.

[12] This view is based on Treitler's view of a "generative system" for chant melodies; see Jeffery, *Past Musical Cultures*, p. 15.

didn't always leave sufficient space—this may be, after all, the first experiment with the idea of writing down text and music—but his intent is clear.

There is another kind of information conveyed in this manuscript; we should mention it briefly, at least to avoid confusion with the systems we will see later that use letters to denote pitch. St. Gall notation uses letters to signify matters of performance practice other than pitch, particularly rhythm and tempo.[13] Over the first neume on the word *Haec*, for example, there is a small *t*, which perhaps stands for *tenere* or *tenete*, meaning "hold the note" or "sing it slowly." Two lines down, over *laetemur*, we see two small *c*'s, which may signify *cito* or *celeriter*, that is, perhaps, "fast." In semiotic terms, these letters function as signs in two different modes—in the symbolic mode, since they stand for words that describe something specific about performance practice, and in the indexical mode, since they require specific physical responses of those who understand the signs— speeding up, slowing down, or some other specific action. This system of rhythmic letters is related to a class of special neumes, such as liquescents, the *quilisma*, and the *pressus*, which also call for specific actions and special modes of performance. For our present purposes, it is important to distinguish between these letters and the letters used elsewhere to signify specific pitches.

Figure 2, from a manuscript identified as Laon 239, represents a slightly later style of notation, the style of notation that Hucbald knew and that we see in the one complete surviving manuscript copy of his treatise.[14]

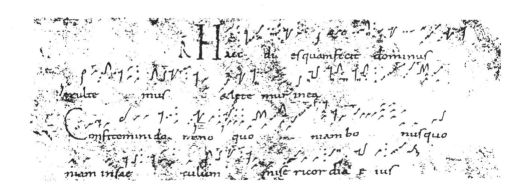

Figure 2

[13] Suñol, *Introducció,* pp. 78–88.
[14] *Paléographie musicale,* vol. 10, plate 103.

This notation has been named "Messine," after the city of Metz, but, more recently, "Lothringian." Note that the neume forms are different from the St. Gall neumes—the *clivis* that begins *Haec dies* now consists of two separate shapes, one over the other, rather than the single cursive stroke of St. Gall notation. Note also that the next neume, a *climacus*, is drawn as three dots aligned vertically, rather than the slash-plus-two-dots in ms St. Gall 359. The next two figures in both manuscripts represent a liquescent *clivis* followed by a *salicus*; those who are interested may want to compare the entire succession of neumes in both sources, to see for themselves that the series of neumes is exactly the same. In designating pitch, Laon 239 is as unhelpful as St. Gall 359. Both manuscripts present a series of signs that function in symbolic mode; if a singer knows the code—that is, if he can translate each sign into a specific neume—he can easily read the series of neumes that fall on each syllable of the text. He still cannot sing the melody correctly, however, any more than he can sight read the melody from the St. Gall manuscript, unless he already knows the "tune." Like the St. Gall manuscripts, this style of notation also utilizes rhythmic letters; if you look carefully, you can see small *t*'s, *c*'s, and *m*'s. Our concern here, however, is the neumes and their lack of clear pitch reference. The first notated manuscripts clearly functioned as memory aids to monks who already knew the melodies; the notation pinpointed for them the location of melismas in reference to syllables of text, and carefully broke each melisma into the proper series of neumes.

Before we look at Hucbald's design for a new kind of notation, let us look, for the sake of comparison, at an example of later notation, in which the problem of denoting specific pitches has been solved by a clever innovation, introducing a radical change into the concept and meaning of notation. Figure 3 is from a

Figure 3

manuscript presently in the Bibliothèque Nationale in Paris, Latin 903, a *Graduale* from the monastery of St. Yrieix (second half, eleventh century).[15]

The manuscript has been said to have been notated in "Aquitanian" notation; it is characterized by a tendency to break up what were cursive figures in St. Gall notation into separate dots, arranged in vertical stacks. More important, notice that there is an entirely new dimension to the notation. In semiotic terms, this notation still functions in symbolic mode; the reader must learn the code in order to read the succession of neumes. He must be aware, for example, that vertical stacks are read downwards, so that the first two figures over the words *"Haec dies"* stand, as in the other manuscripts, for a *clivis* and a *climacus*. But this notation also functions in a new way, in the iconic mode; now the notes are carefully "heightened"—that is, notes of the same pitch are drawn at the same vertical distance above the words, so that the line of notes has the same shape as the melody it represents. What has happened is a crucial change—like our modern notation, this style of notation uses the analogy of the space on the page to represent the ascent and descent of the melody. In other words, the notation now graphs the melody on two simultaneous axes—the horizontal axis denotes relationship with text, just as it did in earlier notation, but now the notator also uses the vertical axis to denote pitch. It is easy for us to imagine how this important new dimension was introduced into notation. Although one can seldom see this in a photograph or photocopy, a dry point guideline runs through the space between lines of text; the text scribe wrote the words on every other line, leaving one dry point line blank between the lines of text. It was a logical step for the music notator to use that line to mark either *C* or *F*, the two notes in the modal scale that have lower half, not whole, steps. The next step is to ink that line, and then add a second line for the other note, either *C* or *F*, and ink them both, often in different colors, usually red and green. From that point, it is another simple and logical step to add two more lines to create the four-line staff, with either a C clef or an F clef, that is familiar from later manuscripts and modern editions of chant. Guido was one of the key figures in the development of staff notation, perhaps his chief weapon in the fight against the monastic system of teaching chant through oral tradition, and certainly the key to his success as an educator.

Hucbald confronted the problem of notating pitch a century before Guido, before there was staff notation, and before the concept of heightened notation had occurred to notators. He agonized over the problem; in his view, the neumes had to be preserved, since they convey important information about other dimensions of the music besides pitch; still, there had to be a way to denote pitch for the reader who had to learn the chant repertory quickly, without the luxury of a long period to follow the monastic system of oral tradition.

Hucbald devised two kinds of notation for denoting pitch. One system works by writing the words of the text on a grid. Figure 4 is Hucbald's illustration of the system.[16]

---

[15] Ibid., vol. 13, plates 152–153.
[16] Brussels 10078/95, fol. 87r.

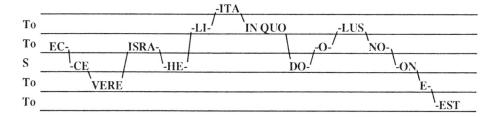

Figure 4

This system functions in iconic mode, since the resulting shape of the line of text mirrors the shape of the melody. Note that the grid is not a staff; only the lines, not the spaces, represent pitch levels, and the abbreviations at the left side mark the intervals between notes. This system is duplicated in several secondary sources, and seems to be better known than the more important system Hucbald actually designed for practical use. While this system has its advantages, it would be quite impractical for a highly melismatic or wide-ranging melody; it is hard to imagine *Haec dies* notated in this system. More important, this system eliminates the neumes altogether, along with the important information about performance they convey, leaving us with a mechanical series of pitches. I mention this system only because it is cited so widely in studies of chant notation, and it tells us something about Hucbald's thought process.

The system that Hucbald actually proposes is more complex. It combines a system of letters to indicate pitch with the familiar neume notation, found in Laon 239. The following is Hucbald's explanation of why such a system is necessary.

> Haec [litterae] ad hanc sunt utilitatem repertae, ut sicut per litteras voces et dictiones verborum recognoscuntur in scripto, ut nullum legentem dubio fallant iudicio; sic per has omne melos adnotatum, etiam sine docente postquam semel cognitae fuerint valeat decantari. Quod his notis quas nunc usus tradidit quaeque pro locorum varietate diversis nihilominus deformantur figuris, quamvis ad aliquid prosint rememorationis subsidium minime potest contingere. Incerto enim semper videntem ducunt vestigio, ut puta si ad subiectam formulam respicias. . . . Primam notulam cum aspexeris quae esse videtur elatior, proferre eam quocumque vocis casu facile poteris. Secundam vero quam pressiorem attendis cum primae copulare quaesieris, quonam modo id facias utrum videlicet uno vel duobus aut certe tribus ab ea elongari debeat punctis nisi auditus ab alio percipias nullatenus sicut a compositore statuta est pernoscere potes. Idem et de ceteris constat. Si vero quas subsequens ratio demonstrabit cordarum notulis eandem

formulam consignatam videris, mox procul dubio qualiter procedit
advertes, hoc utique modo . . .

Al - le - lu - ia.[17]

He points out at some length the lack of pitch reference we have already noted
in early notation styles. Note that he explains the problem as one of
understanding intervals; even when the reader knows that the melody descends,
he cannot know whether it descends a second, a third, or a fourth, if all he has to
guide him is this neumatic notation. If letters denoting pitch are added, however,
the reader can proceed with confidence, knowing exactly what the series of
pitches should be.

At the end of the example is Hucbald's notation for an Alleluia melody.
Although it is difficult to read in the manuscript, there are clearly three elements
to the notation—first the syllables of text, then the neumes, in Lothringian chant
notation, and finally the letters to indicate pitch. Note that the added letters
constitute an additional layer of signs that function in symbolic mode; his design
still does not include an iconic dimension, since the letters do not represent a
similar shape or any other innate similarity to the melody they signify. Until one
is trained in the code, there is nothing to suggest a connection between any
individual letter and the note it represents, or between a series of letters and the
melodic idea signified by the series.

Actually, the system he proposes is more complicated than one might think,
because it is not really a system of letter notation at all. The connection the reader
needs to know between these symbols and the pitches is rather remote; although I
suppose one could learn to read the code efficiently so that the connection
between letter and note could be made readily, the actual system is somewhat
cumbersome. Figure 5 is Hucbald's table of symbols for pitches, as given in
Warren Babb's translation of Hucbald's treatise.[18]

This figure requires some exegesis. To the right is a familiar list of letters for
pitch, the *a-b-c-d* system that became standard; although that column does indeed
appear in the Brussels manuscript, Babb rightly puts those letters in brackets,
since they must have been added by a later hand. Hucbald never mentions that
system of denoting pitches, and always names the notes of the scale in another
way. Hucbald's letters are listed in the second column from the right,
photocopied from the Brussels manuscript. To the left are the Greek string
names, Hucbald's customary names for pitches, taken, as he explains, from a list
of string names in the Lydian mode in Boethius' treatise on music. The shapes
Hucbald calls "letters" are actually his modified version of the list used by
Boethius, who used altered versions of the letters of the Greek alphabet, some

---

[17] Ibid., fol. 90v. For a translation of this passage, see Babb, *Hucbald, Guido, and John,* p. 36.
[18] Babb, *Hucbald, Guido, and John,* p. 38.

extended, some turned sideways or upside-down. For some Greek letters, Hucbald uses lower-case Roman letters.

Nete hyperbolaeon: it has an amplified iota, thus:	Υ	[^a_a]
Paranete hyperbolaeon: [has] an amplified Greek Π		[g]
Trite hyperbolaeon: a simple y		[f]
Nete diezeugmenon: a small N		[e]
Paranete diezeugmenon: a square ω		[d]
Trite diezeugmenon: a simple e		[c]
Paramese: a Greek Π lying on its side		[b]
Nete synemmenon: the same as the paranete diezeugmenon		[d]
Paranete synemmenon: the same as the trite diezeugmenon		[c]
Trite synemmenon: a Greek theta		[bᵇ]
Mese: a simple iota		[a]
Lichanos meson: a simple M		[G]
Parhypate meson: a simple Greek P [Rho]		[F]
Hypate meson: a simple Greek sigma		[E]
Lichanos hypaton: a simple digamma		[D]
Parhypate hypaton: a simple beta		[C]
Hypate hypaton: a simple gamma		[B]
Proslambanomenos: an upright daseia		[A]

Figure 5

In other words, this is not really a system of letter notation, because the chain of symbolic functions is quite involved, and leads finally to words, not letters, that stand for pitches. When we see in Hucbald's example the letters *i, m, p, c,* and *f,* we might conclude that he is proposing an alternative letter system that we can translate into the more familiar *a-b-c-d* system. But each letter actually functions as a complex chain of symbols—*i* is "iota," which signifies *mese,* which in turn signifies *a*; what looks like *p* is actually "rho," which signifies *paryhypate meson,* which stands for *g*, and so forth. It is probably more accurate to classify this system as a system of verbal symbols, rather than, as he states, a system of letter symbols. Modern readers perhaps find his system more cumbersome than it really is, because of our difficulty with the Greek string names, a standard system of pitch reference in Hucbald's time. We cannot help wondering if his system would not have been more successful if, instead of the Greek string names, he had used the *a-b-c-d* system.

As it turns out, letter systems similar to this one continued to be used for some time; a recent study of medieval letter notations lists nearly sixty manuscripts, not theoretical works but practical sources from the tenth through the thirteenth centuries, containing chant and polyphony notated in some variety of letter notation.[19] The more successful systems use the *a-b-c-d* system and write text, pitch letters, and neumes in three parallel horizontal lines. That system is still cumbersome, but we can imagine that a singer could actually learn the system and read the three lines at once, at least after being somewhat familiar with the melody, if not at first sight.

Having explained how his pitch letters work, Hucbald goes on to explain why he insists on keeping the neumes in his system, even though they are no help with the question of pitch. Here is his explanation of the advantages of keeping the neumatic notation.

> Hae tamen consuetudinariae notae non omnino habentur non necessariae, quippe cum et ad tarditatem seu celeritatem cantilenae, et ubi tremulam sonus contineat vocem, vel qualiter ipsi soni iungantur in unum vel distinguantur ab invicem, ubi quoque claudantur inferius vel superius pro ratione quarumdam litterarum, quorum nil omnino hae artificales notae valent ostendere, admodum censentur proficuae. Quapropter si super aut circa has per singulos ptongos eaedem litterulae quas pro notis musicis accipimus apponantur, perfecte ac sine ullo errore indaginem veritatis liquebit inspicere, cum hae quanto elatius quantove pressius vox quaeque feratur insinuent; illae vero supradictas varietates sine quibus rata non texitur cantilena menti certius figant.[20]

Note that Hucbald values the neumes for what they signify about performance practice. He singles out those neumes, like the *quilisma* and liquescents, that signify subtleties of performance beyond pitch. In his conclusion, he expresses great satisfaction with his system; the singer reading his notation will be absolutely sure of pitch, a matter not dealt with in previous notation, and still know all the proper subtleties that will turn a series of pitches into a true chant melody.

---

[19] Alma Santosuosso, *Letter Notations in the Middle Ages* (Ottawa, 1989).

[20] Brussels 10078/95, fol. 90v. For a translation of this passage, see Babb, *Hucbald, Guido, and John*, p. 37.

In summary, then, Hucbald proposed a two-level system of music notation, combining a series of letters—actually, symbols for the Greek names for the strings of the lyre—to signify pitch, and the traditional neumes to signify everything else. He worked long and hard to design a way to provide the different kinds of information the reader needs in order to re-create a chant melody. We cannot help reflecting on the short life of the system he designed. Had he lived a century later, and thought more about the possibilities of iconic symbols, he might have been as famous as Guido. He was on the brink of designing a system of iconic symbols with his system of syllables on a grid (as in figure 4), but he apparently never considered the possibility of combining the concept of melodic shape with the neumatic notation that he knew he had to preserve.

Finally, Hucbald's design for music notation is a perfect illustration of the way the medieval mind works. His solution to the vexing problem he had been called upon to solve is not to create a completely new system of signs that can clearly signify chant melodies to the new students in the cathedral schools. Instead, he felt required to keep the symbolic signs of what was perhaps earlier chant notation, and decided to add another layer of complex signs, also functioning in symbolic mode, to signify the dimension missing from the neume notation of his day. His design, with its parallel levels of sign systems, functions, I believe, by adding a layer to the existing system, in the same way as both tropes and early polyphony added a new layer to the traditional chant, and in the same way as medieval thought usually advanced, by adding a new layer of marginal glosses and commentary. No notation system can convey everything about music, but the efforts of monks like Hucbald to design a system that would convey as much information as possible about the glories of the immense chant repertory eventually succeeded. Because of them, instruction in the art of singing chant moved successfully out of the monastery into the new world of the cathedral school, and eventually into the university.

UNIVERSITY OF SOUTHERN CALIFORNIA

Seeing Through Hearing: The Construct of the *Cithara* in Medieval Biblical Interpretation

by Nancy van Deusen

> I will incline my ear to a parable,
> > *open up* my proposition to the psalterium
> (Inclinabo in parabolam aurem meam,
> > *aperiam* in psalterio propositionem meam)

which has been translated in the Luther Bible as:

> Ich will mein Ohr einem Spruche neigen,
> > bei Harfenklang mein *Rätsel* lösen

is the inviting opening line to a subject presented in Psalm 49 (Vulgate 48), one of the psalms which periodically throughout the Psalter present the special function and properties of the stringed music instruments, the *psalterium* and the *cithara*. Three linguistic renderings of this passage have been included here, since each brings to the context a different aspect within a psalm that appears to require some explication. Augustine, in one of his commentaries or sermons on this psalm visibly warms up to the task of providing an interpretation. He, as teacher, bishop, hence, shepherd of the sheep, begins his expository treatment with a series of questions, meant, certainly, to engage his flock, perhaps in the cathedral of Hippo, on a Sunday morning. *Quis, quis,* and *qui*—in other words, one question after another—follow as he writes: *Inclinabo in parabolam aurem meam, aperiam in psalterio propositionem meam.* I will incline my ear to a parable, open my proposition onto the *psalterium*. A parable is also a *figura*, the Latin translation of the Greek *schema*, a multifaceted term that included the conceptual nuances of, and therefore could be translated into Latin vocabulary indicating, design, rhythm, number, letter, character, type, instrument, and gesture—to name some of many nuances of this multivalent term. In this short statement from the Psalms, therefore, the concept of *schema-figura* would occur explicitly three times, with an additional fourth shade of meaning to an educated person in late antiquity, such as Augustine, or an early-medieval biblical commentator, such as sixth-century Cassiodorus. *Parable* (*parabolos*), *psalterium, proposition* (*propositio*) are all *schemata*, and *inclinabo*—I will incline—presents the additional gestural significance of *figura.* So we have a tightly-woven conceptual cluster in which, from the very beginning of this psalm, every word is carefully chosen and utterly significant. A quotation from Augustine's Latin commentary provides an example:

> What is to be understood by this, as we consider the very *heart*, and not only the superficial *sound* of the tongue, that is, the inner life of a human being? *(quis est hic cuius meditatio cordis loquitur*

intelligentiam, ut non sit in sola superficie labiorum, sed hominis interiora possideat?) What should be heard? and what is said? Much is indeed said that is *not* heard. [Augustine is surely speaking from his experience as teacher.] Who are those to whom much is spoken but nothing heard? Is it not to those who really desire to hear what the Psalmist writes, as he begins: I will explain my riddle on the *psalterium*. This is to say, through a body. Here is a corporeal presentation of what has strictly to do with the mind (*quod est iam loqui per corpus, sic enim utitur anima corpore*). In just such a way, that is, to bring out meaning, the *citharista* uses the *body* of his instrument, saying: I will incline my ear to a parable.

You ask, what parable? We see now through a glass darkly, while we are in this body, travelling on and on to be, one day, with the Lord, writes the Apostle Paul. In this body, our vision is never face to face. What is seen is through parables, enigmas, and similitudes. In what respect do we see as if enigmatically? An enigma is an obscure parable that is difficult to understand. Because of the corruption of this body, we see in part, we understand in part (*quamdiu per corruptibilitatem carnis huius videmus, ex parte videmus.*)[1]

We see through a glass darkly, writes Augustine, quoting Paul in one of the most famous passages of the Apostle's combined epistles.[2] But there is an implication that we hear perfectly well. Sight, face to face, is impossible, but clarity of the auditory faculty is implied. The implication, granted, among many, in this passage that is rich in allegorical significance, is that even to the

---

[1] *Enarrationes in Psalmos* 48.5 (Corpus christianorum series latina, vols. 38–40, ed. Eligius Dekkers and Johannes Fraipont [Turnhout, 1956], 38.554), the entire Latin context: *Inclinabo in parabolam aurem meam, aperiam in psalterio propositionem meam.* Quis est hic cuius meditatio cordis loquitur intellegentiam, ut non sit in sola superficie labiorum, sed hominis interiora possideat? Quis est iste qui audit, et sic loquitur? Multi enim loquuntur quod non audiunt? Qui sunt qui loquuntur quod non audiunt? . . . Iste ergo qui et auditor volebat esse et dictor, qui tibi loquitur, antequam diceret: *Aperiam in psalterio propositionem meam*, quod est iam loqui per corpus, sic enim utitur anima corpore, quomodo utitur citharista psalterio, dixit: *Inclinabo in parabolam aurem meam*. Antequam loquar tibi, inquit, per corpus, antequam psalterium sonet, primo ego *inclinabo in parabolam aurem meam*, id est, audiam quid tibi dicam. Et quare: *in parabolam*? Quia *videmus nunc per speculum in aenigmate*, sicut dicit apostolus. *Quamdiu sumus in hoc corpore, peregrinamur a Domino.* Quia nondum est illa visio nostra facie ad faciem, ubi iam non sint aenigmata et similitudines. Quidquid modo intellegimus, per aenigmata conspicimus. Aenigma est obscura parabola quae difficile intellegitur. Quantumvis excolat homo cor suum, et ad interiora intellegenda refugiat, quamdiu per corruptibilitatem carnis huius videmus, ex parte videmus. Assumta autem incorruptione in resurrectione mortuorum, cum apparuerit Filius hominis iudicaturus vivos et mortuos, tunc videbitur Filius hominis, qui primo iudicatus est, iudicans, discernens malos a bonis, ponens malos a sinistris, bonos a dextris.

[2] 1 Corinthians 13:12: For now we see through a glass, darkly; but then face to face: now I know in part; but then shall I know even as also I am known.

imagination, the inner cognitive deliberation of a music instrument with its body is enough to transmit a message that is, so to speak, loud, clear, and plain.

I have chosen not to go further into tropological implications of Augustine's interpretation, namely, that bodily activity, or deeds related to the performance of daily life show—as the *cithara* player demonstrates his musicality when he performs—an inner landscape, that is, the propensity, motivation, and capability in terms of inner property and training, to play this instrument. One demonstrates one's otherwise invisible character by one's affinity to an instrument, as well as what is actually played upon that instrument, and the degree of competence with which it is played. It is a clear case of the inner character or *figura* of the player corresponding to the outer *figura* of the instrument, in which, ideally, the outer corresponds to the inner. All this would have been clear to Augustine's congregation, since he had used the same analogy many times before in his sermons.[3] I will rather focus upon two aspects, drawing attention again to the fact that "seen" things may be ambiguous, as if through a glass darkly, but sound is not ambiguous, hence, in part at least, a musical analogy to what is difficult to comprehend. Secondly, Augustine presents a comparison of *figura/corpus*, a distinction which, again, would have been perfectly clear to his fourth-century audience. Nearly every significant noun in Augustine's short commentary is a *figura*, as we have seen, with the one exception of *corpus*, which derives from another Greek term, therefore an entirely different concept. This distinction is also important to Cassiodorus, whose commentary is shorter, since he knew and was building upon Augustine's *Sermones*, but also because Augustine was a bishop charged with pastoral care—who also took this charge very seriously indeed—and Cassiodorus was a scholar-teacher, more interested in identifying, distinguishing and demonstrating how *figurae* function. I quote from Cassiodorus' commentary on the same passage:

"I will incline my ear to a similitude, open up my proposition upon the psalterium." We notice that Cassiodorus uses *similitude* instead of *parable*, indicating, in this way, the fact that more than one translation of the Psalms were available to both Augustine and Cassiodorus. Both of them could have chosen the Latin translation with just the right nuance for the purpose at hand, and this, apparently is exactly what each did. The psalm commentator, in tight, succinct Latin, which strikes the reader as academic in that it, for the most part, defines terms, interprets the passage as:

> In eloquent language two predicates are set forth as significant, now [the question is] in what way do these form precepts applicable to human experience in general. The phrase, to incline the ear, indicates

---

[3] Augustine comments on the *psalterium/cithara* in his comments on Psalms 32.2.5 (CCSL 38.250), 48.5 (CCSL 38.554), 56.16 (CCSL 39.705), 67.34 (CCSL 39.893), 70.11 (CCSL 39.970),80.4 (CCSL 39.1122), 91.5 (CCSL 39.1282), 97.1 (CCSL 39.1372), 103.17 (CCSL 40.1534), 107.1 (CCSL 40.1583), 136.1 (CCSL 40.1965), 143.16 (CCSL 40.2084), 146.1 (CCSL 40.2121), 150 (CCSL 40.2195). Cf. van Deusen, "The *Cithara* as *Symbolum*: Augustine *vs.* Cassiodorus on the Subject of Musical Instruments," in *The Harp and the Soul: Essays in Medieval Music* (Lewiston, NY, 1989), pp. 201-255.

what can be recognized, encouraging popular devotion. [I think we can translate this "popular devotion": *populus devotus impleret*.] But here a similitude is proposed. A similitude is an imitation of the real thing, given to us as an example that we should follow in the emulation of our Lord who is present with us (*Similitudo* enim rei verae *imitatio* est, ut quod nobis ad *exemplum* datum est, devota *aemulatione* [Domino praestante] faciamus.) And so the precept of salvation is available to all. The gracious institutor, Christ himself, extends the invitation to make open and apparent, as he states his proposition by means of the *psalterium*, that is, a declaration of his own divine holiness by a property of his own body. The Lord teaches, not only verbally, but by very example. The *psalterium* then—as we have often said, is a *similitude* which *is* itself the body of Christ, for just as the *psalterium* sounds from the upper part, so resounded the celestial mandate of the Lord's incarnation (*Psalterium* quippe [ut saepe diximus] corporis Domini decora similitudo est; nam sicut psalterium de summo sonat, ita et incarnatio Domini caelestia mandata concelebrat).[4]

It is interesting to note both similarities and differences in the two relatively short commentaries. What is similar is that neither commentator appears to distinguish between *psalterium* and *cithara* in this case. "A generalized many-stringed instrument" will do for their commentaries this time. Augustine's comments emphasize the ambiguity, difficulty, obscurity, and mystery of the *parabolum*. *Psalterium*, however, is not so interesting to him. One has the impression that had the psalm writer not included the *psalterium* in this psalm, Augustine would not have brought the instrument up, since he concentrates on what is obscure, not plain. A parable is an enigma, difficult to grasp, and something other than that which is viewed "face to face." Cassiodorus' *similitude* of the *psalterium* is quite the opposite: a similitude opens up, declares and clarifies what is intended by the example. Cassiodorus' instrument is a similitude, which would have been to Cassiodorus—as well as his medieval readership—a tautology for purposes of reinforcement: a *figura* (*psalterium*) is a *figura* (*similitudo*). Nobody would have missed this.

The *psalterium* gives forth sound, has a body, but this is not as important as the fact that the music instrument teaches by example, not word. The *psalterium*, in the manner of a simile, makes known the fact of the incarnation, the bodily form of God. Further, whereas Augustine's discussion in no way demands the use of a specifically *musical* instrument, Cassiodorus' interpretation depends upon, and stresses the unique value of the *psalterium*. The musical instrument is qualified in a special way to serve as *exemplum*, bringing that example face to face, providing a physical statement of a musical reality, though conceived by the imagination, just as a similitude provides a verbal example of an intellectual reality. Instrument, in this case, is a *figura*, both directly, as a translation of the

---

4 Magni Aurelii Cassiodori, *Expositio psalmorum*, 2 vols. (Corpus christianorum series latina, vols. 97–8, ed. M. Adriaen [Turnhout, 1958]), 1.433 f.

Greek term *schema*, but also in its function as a *similitude*, which is also a *figura*. But Cassiodorus' stringed musical instrument was also a *body*. The main distinction to be made between both of these authors is that whereas Augustine sets forth a parable—an enigma—*per corpus*, Cassiodorus' explanation is *through* a *figura*—a similitude. Augustine finds obscurity *in* the *parabolum* (a *figura*); Cassiodorus, on the other hand, finds clarity *through* the *figura* of the similitude.

Three observations can be made at this point. One notices first the sheer importance of the *psalterium, cithara, citharista figura*. Both commentaries, and many, many more on the Psalms, rest on the significance that this specifically musical figure brings to the Psalms. Secondly, nearly everything mentioned in both commentaries was considered to be a *figura*, that is, was made or artifacted with a specific purpose in mind, was delineatory, significant, indicative as an outer design indicates internal connection, substance, or property. Further, a *figura* was instantly recognizable, partially because of its sharply-delineated qualities, and partly because *figurae*—such as numbers, letters of the alphabet, for example—became conventional. This brings up perhaps the most important attribute of *figurae*, namely, that they occur in groups or series of varied and diverse figures. Each one belongs to an entire system of figures, by which means they are compared and find meaning. Finally, once the nature and identity of these *figurae* had been established, a distinction could be made between *figura* and *corpus*. And this is exactly what eventually happened.

First, a *corpus* is a composite, not an individual *figura*. Eventually, by the time Cassiodorus has finished actually commenting on all of the psalms that mention the *cithara*, build upon it as a *figura*, or make the *cithara* an important aspect of their structure, the psalm commentator has managed to bring together a *body* of attributes for this stringed instrument. The entire assemblage of component interpretative modules for the *cithara*, together, forms a powerful interpretative device. Each treatment of the *cithara* in Cassiodorus' commentary supplies a separate component part of this exegetical *corpus*. But this is a particular type of instrument. Cassiodorus' composite *cithara* is an instrument that provokes the heart, rather than caressing the ear. It is not necessary for this exegetical *cithara* to actually produce sound.

Cassiodorus' corpus of distinctive qualities for the *cithara*, however, though comprehensive, was a body that required a *written* elucidation of all component parts so that, eventually, in reading, one could refer back and forth and collect them into one complete, meaningful whole. The component parts, one by one, could not be related to each other unless one had carefully read through Cassiodorus' complete *Expositio psalmorum*, writing down each aspect of the *cithara/psalterium* construct as one went along, until, at the end, one had the entire corpus of what Cassiodorus had to say on the subject. A medieval person interested in the *cithara* could have done this too, since the *Expositio psalmorum* was transmitted together as a *Liber*, whereas Augustine's *Sermones* or *enarrationes* on the Psalms were not. This point is important for the perception of the totality of all of these *figurae* of the *cithara* within one cogent body. My point here is that this totality, the complete *corpus* of attributes of the *cithara* was not one that snapped out of Cassiodorus' writing, but rather required steady perseverant study, then, upon reflection, seeing the point, and bringing, by comparison, all of the components together. And it required a selective process in which one culled all of the *cithara/psalterium figurae* from the hundreds of other *figurae* that Cassiodorus names, classifies, and comments upon in his energetically pedantic manner. This is the process of extracting a *corpus* from the immensity of varied and diverse figures.

Cassiodorus' body of the *ecclesia*, or "world-wide Christian church," is another case in point, another example of this structure, which will, perhaps, illuminate the distinction made here. The *ecclesia*, according to Cassiodorus, was a circle or a crown, on which each Christian believer—from all parts of the world, and all periods of history—formed a decoration.[5] Not only the separate features but, especially, the totality of the crown, therefore had to be imagined, and one had to have both a considerable acquaintance with the biblical scriptures, as well as a good imagination to do so. In Cassiodorus' body—a totality composed of figures—the component parts, which he meticulously identifies, take precedence, each one in its figural distinction. The totality, or body, is an abstraction.

It seems to me that one finds this feature of figural importance, with a correspondent disinclination toward the totality, or *corpus*, both in further medieval exegesis of the *cithara*, as well as in illuminations of the Psalter. In the *Allegorae in sacram scripturam*, a twelfth-century dictionary of allegorical figures, believed for many years to have been written by Rhabanus Maurus, the *cithara* forms a *figura* for these attributes: "*Cithara est recta operatio, ut in Psalmus 'Confitemini Domino in cithara'* [*Cithara* is right operation, as in the Psalms, I will place my confidence in the Lord on the *cithara*], *vita activa, id est dulcidinem contemplatione cum actione* [*vita activa*, that is sweet contemplation with action—in other words, concord]; *mortificatio carnis, laetitia, ut in Job*: Versa est in luctum cithara mea,"[6] sweet consolation (suavitas consolationis).[7]

---

5 Cassiodorus, *Expositio psalmorum*, 1.183.4: "Haec erat corona capitis, hoc regale diadema, quod non ornaret impositum sed de Christo Domino potius ornaretur. In hac enim corona et totius mundi circulum merito poterimus advertere; in quo generalis significatur Ecclesia. Quod schema dicitur characterismos, id est informatio vel descriptio, quae sive rem absentem sive personam spiritalibus occulis subministrat."

6 Cf. *Planctus Davidis.*

7 Pseudo-Rhabanus Maurus, *Allegoriae in sacram scripturam*, PL 112, col. 897: *Cithara* is "recta operatio, vita activa, mortificatio carnis, laetitia," as in "dolorem laetitia mea, suavitas consolationis." It should be noted that all of these attributes for the *cithara* usually come together; if one has one, one has all, and all together form a balanced composite containing order in serene organization and appropriateness, the practical in combination with "sweet contemplation," discipline, joy, and loving kindness. All are obviously positive; all are designated by the *figura* of the *cithara* itself. In the following quotation, from Flodoardus Remensis, again, a stringed instrument brings together the varied and diverse in abstract terms, referring, as well, to the influence of Cassiodorus: "Cassiodorus . . . Davidis carmina tractat, clarificans supero modulata poemata nablo illustransque pias divino lumine turmas" (*De triumphis Christi sanctorumque apud Italos* X.12, PL 135, 771C).

The figure of the *cithara* matches, in all of these instances, inner attributes. Yet, this *figura* is imaginary, or remains figural. It is important that a stringed instrument be imagined, but exact specifications are not required, or even desired. Further, in every case, Pseudo-Rhabanus Maurus' inner attributes of the outer *cithara figura* are obviously positive: confidence, the balanced concord of contemplation and action, self-discipline, sweet consolation. The *cithara* is a figure, not a body, and indicates, one way or the other, the otherwise unseen substance of inner spiritual righteousness for a reading public of late antiquity and well into the Middle Ages. These two attributes, the figural and the positive, accompany inclusion of the *cithara* within a medieval psalm iconographical tradition as well.

I am interested in what I consider to be a break with this firmly-established *psalterium/cithara* tradition. This rupture takes place in both spheres, that is, the *cithara* as *figura*—spare, delineated, articulated in terms of conjunctive lines, and making distinct what can be articulated as diverse inner attributes. The second rupture is with the long tradition of the *cithara* as delineator of essentially positive aspects—the inner spiritual substances of the righteous person. These two breaks with what can be seen to be a long exegetical-iconographical past can be attributed to very specific causes, and thus give us an idea of how philosophical concepts achieve results, as well as the connection between concept and depiction. First, let us take on the first break with the past, that is, *cithara* consciously presented as *body*, not as *figura*.

Although Aristotle's treatise, *De anima, Concerning the Soul* had, during the course of the twelfth and early thirteenth centuries, been previously translated into Latin, it was William of Moerbeke's translation of the 1260s that proved, quite suddenly, to be tremendously influential. There are 268 extant copies of this particular treatise even today.[8] Reasons for this surge of interest and influence are not hard to find: *De anima* summarizes what had been believed concerning *figurae* in the Middle Ages; the treatise deals with the most important connections made by the *Phaedo* of Plato, which, translated in the mid-twelfth century, was also, around 1275 to the mid-thirteenth century, sensational, inasmuch as it was the first complete Platonic treatise to be available in Latin, thus accessible to the Latin-reading public. The two treatises, the *Phaedo* and *De anima,* deal with the nature of the soul, and it is a testimony to the importance of the late-thirteenth-century Latin translation of *De anima* that the treatise is still today customarily referred to by its Latin, not vernacular, title.

---

[8] For medieval translations of *De anima*, see Bernard G. Dod, "Aristoteles latinus," in *The Cambridge History of Later Medieval Philosophy,* ed. Norman Kretzmann, Anthony Kenny, Jan Pinborg, with Eleanore Stump (Cambridge, 1982), pp. 46 f. *De anima* was translated by James of Venice, presumably between 1125–50, by Michael Scot c. 1220–35, and by William of Moerbeke before 1268. In addition Averroes' great commentary would have been available, translated c. 1220–35 by Michael Scot.

Example 1. Here, delineatory lines indicating *figurae* are presented on
several levels, as well as the topos of "varied and diverse figures";
further, the use of the *psalterium-cithara figura* is positive. (Paris,
Bibliothèque nationale, f. lat. 1, fol. 215r [c. 850])

De anima has this to say about the soul and its relationship, not to a *figura*, but to the *body*. Aristotle writes that the facts "regarding the soul are much the same as those relating to figures," a statement which would have connected what he had to say next to the medieval traditional discussion of figures; but then there is a profound deviation from the familiar distinction made between the unseen, which was soulish, and the seen, the *figure*. Figures, here, are not important, rather, body. So writes Aristotle:

> Probably it is better not to say that the soul pities, or learns, or thinks, but to say rather that the soul is the *instrument* whereby man does these things.[9]

The soul causes the body's movement. Aristotle then summarizes what he has to say about soul and body, using music instruments as *exempla*: "The soul is the cause and first principle of the living body," and finally, "*all natural bodies are instruments of the soul*."[10] There is, accordingly, to Aristotle, an ensoulment that takes place; the body, as instrument, becomes *ensouled*.

And there is dynamism—sound and movement. Aristotle becomes very specific about musical instruments as examples of ensouled bodies:

---

[9] Aristotle, *De anima*, Engl. trans. W. S. Hett, Loeb Classical Library 288 (Cambridge, Mass., 1936, repr. 1986), p. 47.

[10] Aristotle's succession of thought is the following: p. 41 f.: ". . . on the contrary, the soul causes the body's movement . . . Men associate the soul with and place it in the body, without specifying why this is so, and how the body is conditioned; and yet this would seem to be essential. For it is by this association that the one acts and the other is acted upon, that the one moves and the other is moved; and no such mutual relation is found in haphazard combinations. But these thinkers only try to explain what is the nature of the soul, without adding any details about the body which is to receive it . . . every body has its own peculiar shape or form [his example is a flute]; each craft must employ its own tools and each soul its own body"; p. 61 (elements, pairs of contraries): "And if we are to construct the soul out of the elements, it is unnecessary that it should be composed of all the elements; for only one of a pair of contraries is needed to discern both itself and its opposite"; p. 63 (*operatio*): "But since knowing, perceiving, and the forming of opinions are operations of the soul, besides desiring, wishing, and the appetites in general, and again since movement in space is induced in living creatures by the soul, besides growth, maturity, and decay, does each of these belong to the soul as a whole? Do we think, perceive, and do or suffer everything else with the whole soul, or do some functions belong to one part and others to another? Does life reside in one or several or all of these parts or is something else the cause of it? Some say that the soul has parts, and thinks with one part, and desires with another. In this case what is it which holds the soul together if it naturally consists of parts? Certainly not the body: on the contrary the soul seems rather to hold the body together; at any rate when the soul is gone the body dissolves into air and decays"; p. 67: "Bodies seem to be pre-eminently substances, and most particularly those which are of natural origin; for these are the sources from which the rest are derived"; p. 83: "The facts regarding the soul are much the same as those relating to figures (σχημάτων), for both in figures and in things which possess soul, the earlier type always exists potentially in that which follows"; p. 89: "*all natural bodies are instruments of the soul*"; p. 107 (sensation): ". . . For vision occurs when the sensitive faculty is acted upon; as it cannot be acted upon by the actual colour which is seen, there only remains the medium to act on it, so that some medium must exist . . ."

> Now sound in actuality is always of something, and against something, and in something. For it is a blow that produces it. . . The sounding thing, then, sounds against something, and no blow can come about without movement. But, as we said, sound is the striking of the right kind of things. For wool, if struck, would make no sound, but bronze would, and the things that are smooth and hollow . . . Now voice is a kind of sound of an ensouled thing.[11]

Aristotle's "ensouled instruments" in which actuality is of something, against something, and in something, in short, the range of idiophones and membranophones which he then, in the *De anima*, proceeds to describe, can be found in the early fourteenth century *Roman de Fauvel* (see below). (Notice that in every case, strikers are distinct from what is struck; motion, as well, pervades the depiction, and the stringed instrument is Aristotle's "analogy," that is, the *figura* of the *lira* in contrast to instruments struck, smooth, and hollow.)

Example 2. Paris, Bibliothèque nationale f. fr. 146, fol. 34r

---

[11] Aristotle, *De anima*, trans., with an Introduction and Notes, Hugh Lawson-Tancred (London 1986), pp. 176–8. (I have chosen what I have considered to be the best translation of a selected passage from the two English translations indicated.) Judith Becker has observed to me that this is true, as well, of Indian music. For struck vs. non-struck parallels with other world music civilizations see Arnold Bake, "The Music of India," in *Ancient and Oriental Music*, New Oxford History of Music 1, ed. Egon Wellesz (London, 1957), pp. 195–199; Terry Ellingson, *The Mandala of Sound: Concepts and Sound Structures in Tibetan Ritual Music* (Ph.D. Dissertation, University of Michigan, 1979) pp. 148–165; Lewis Rowell, *Music and Musical Thought in Early India* (Chicago, 1992), pp. 43–55.

The connection made by Aristotle between soul and body, sound—abstractly, the nature of sound—and instrument, is not an earlier medieval connection. Although this connection is also between one, and possibly even two, unseen but imaginable entities, soul, and the body of an imagined instrument, it is a relationship which is easier to grasp instantaneously. Further, it is a relationship that is supported conceptually by the equivalence presented in Plato's *Phaedo*, of the relationship of soul-body of a stringed instrument-intrinsic harmony. The *Phaedo* was translated from Greek into Latin in 1157 by Henricus Aristippus, and very quickly found its way into collections that would subsequently form two of the major libraries of Europe, namely the Sorbonne and the Vatican. By 1300, the *Phaedo* was available for the Latin-reading public, and, subsequently, if we are to believe Petrarch, was known, in fact, read. It was, after all, the first Platonic work to be available in its entirety to the Latin-reading public since antiquity, and caused quite a stir.[12]

What difference does this make? What change does this effect in *cithara* interpretation? First, the emphasis is shifted from individual *figura* to a composite whole, an equation of soul (unseen) with body (as instrument of the soul, seen, making sound, demonstrating motion), in which component parts are kept in tension with one another to produce, as Plato states, a harmony of diverse elements. It is the distinction, which Aristotle also makes, between an analogy, a *figura*, (that is, the *lira*), and an actual resonating body. The body is also much more of a conceivable totality. Secondly, the imaginary construct of the *cithara* has been separated from its firm position within psalm exegesis where it had remained for so many hundreds of years. This is a major change in orientation— dramatic, if one compares psalm miniatures with the *charivari*, or *charactere variarae* of the *Roman de Fauvel*, in which all of Aristotle's sound-producing instruments are present, combined with varied and diverse inner properties indicated outwardly by masks.

I am, of course, making an interpretation here, that is, that *cithara, psalterium, lyra* (which is the term Plato uses in the *Phaedo*, and only in the *Phaedo*), and a bowed vielle-type all belong to the same generalized stringed-instrument analogy, which serves as outer delineatory figure. It is clear that each one of these instruments has a distinct organological character. But I do not believe that this is the medieval point. Frame, strings, the fact that there are several strings, and that they are all separate, the composite nature of the instrument, the fact that the instrument is external, can be seen, and is indisputably a *music* instrument are all important to the connection that is made. But, as a myth with broad, general, defining outlines, details in this stringed instrument vary. The *charivari* contains, however, in addition to the spectrum of Aristotle's instruments, a scurrilous quality, an undercurrent of furtiveness, simultaneously, obvious *disharmony* in terms of faces wagging, feet kicking, hands moving in contrary directions. There are actually pairs of contraries, a concept that is also found in Aristotle's treatise. The percussion produces an impression of impetuosity, of imagined motion—

---

[12] Cf. Nancy van Deusen, "The Harp and the Soul: The Image of the Harp and Trecento Reception of Plato's *Phaedo*," in *The Harp and the Soul: Essays in Medieval Music*, pp. 384 ff.

certainly not stability. *Cithara, psalterium,* Plato's *lyra* all indicate *harmony,* the well-proportioned and appropriate balance of strings within frame delineating unseen but obvious qualities such as righteousness and health. Outer sound corresponds to inner soundness. For Plato, it is a healthy, harmonious soul that is represented by the *lyra,* when all of its strings are in proper tension. The change must have been jarring for viewers accustomed to the *cithara* as convention.

Example 3 is an altarpiece that must have been even more of a shock. Grünewald's altarpiece combines the *cithara* exegetical tradition with what I take to be the devil playing a stringed instrument.[13] (Examples 3–6 from Grünewald, Altarpiece at Isenheim [c. 1512–1516], commissioned by Abbot Guido Guersi, now installed in Colmar's Unterlinden Museum.)

Using the language of Aristotle's *De anima,* the musical instrument is ensouled. The stringed instrument becomes enlivened by an obviously negative inner force. Not only is the stringed instrument ensouled by the devil himself. If one keeps in mind that, in Plato's *Phaedo,* the human soul is represented by a stringed instrument, then the musical instrument, ensouled by this ghoulish being, could be, as well, the human soul. The contrast would surely have occurred to a sixteenth-century viewer that God does *not* play the human soul as an instrument, but rather, as Cassiodorus states, invites the human race to salvation, using his own body, and although the related subjects of attraction, invitation—even seduction—of the soul by God can be found, particularly in Song of Songs commentaries throughout the Middle Ages, God never actually *plays* upon the soul.[14] What we have here is a remarkable insight and a stunning break with the past, as well as an indication of just how much theology Grünewald himself knew—actually a good deal. The visual-conceptual *topos* of the devil with his violin, traceable to Stravinsky's *L'Histoire du soldat,* would follow.

---

[13] The Isenheim Altarpiece, understandably, has attracted much attention; however, in spite of focus on nearly every other aspect of the altarpiece, no one has mentioned this stunning break with a powerful exegetical tradition. For example, recently, Ruth Mellinkoff, in going carefully over every aspect of the altarpiece, including the chamber pot—which is important to her argument—did not consider the significance of any of the music instruments. Her point, however, that Lucifer is depicted as playing a viol is an important one, on which I have built my argument. Cf. *The Devil at Isenheim: Reflections of Popular Belief in Grünewald's Altarpiece,* California Studies in the History of Art, ed. James Marrow (Berkeley, Los Angeles, 1988). Andrée Hayum in *God's Medicine and the Painter's Vision* (Princeton, 1989) esp. pp. 44 ff. makes general comments concerning the healing effects of music, but gives no indication of the long and important *cithara* expository tradition, or cognizance that the stringed instrument in Plato's *Phaedo* relates soul with attunement that is both healthy and harmonious. Both of these facets, I believe, would have immediately sprung to mind for a contemporary audience.

[14] Since this would violate man's free will, God does not manipulate. Only rarely does God "play the strings" of a soul already repentant, who has already turned to God. See Mechthild of Magdeburg, *The Flowing Light of the Divinity,* trans. Christine Mesch Galvianai, ed. Susan Clark (New York, 1991), p. 142. I am grateful to Professor Marcia L. Colish for directing me to this reference.

Example 3

Example 4

Example 5

Example 6

What change has actually occurred between the *cithara* as a *figura* in medieval psalm exegesis, and a stringed instrument as an ensouled *body*? Among other factors, it is a new "dynamism," I believe, coming from a source that can be identified, then filled with new possibilities for pictorial representation. *Figurae* are essentially static, and bonded to words, since *figurae* are also letters of the alphabet. Aristotle's concept of the musical instrument as an ensouled body is dynamic. Aristotle's seen, enlivened instrumental bodies, in contrast to Augustine's *cithara*, can, visually, be heard.

CLAREMONT GRADUATE UNIVERSITY

Weeds in the Garden of Eloquence:
The Proliferation of Rhetorical Figures

by Carol Dana Lanham

Thus much of Eloquution in tropes and figures:
in al which obserue this one lesson, the more the better.
—Abraham Fraunce, *The Arcadian Rhetorike* (1588)

I was lured into the garden of eloquence, almost against my will, first by my long-standing interest in epistolary theory, the rules for writing letters that developed into the *ars dictaminis* shortly after 1100. The earliest known handbook to include such instruction is a general rhetoric text by Alberic of Monte Cassino, composed perhaps around 1075; manuscripts title it *Flores rhetorici* ("flowers of rhetoric") or *Radii dictaminum* (something like "bright rays of composition"), and the bulk of it is devoted to style, particularly a series of rhetorical figures.[1] In tracing Alberic's sources, I was struck by the paradox that while he relied heavily on Isidore's catalogs of figures in books 1 and 2 of the *Etymologies*, Alberic seemed deliberately to avoid using Isidore's terminology and definitions, preferring to invent new terms for the same figures even when he used traditional examples to illustrate them.

Next, I began working with my husband to prepare a second edition of his *Handlist of Rhetorical Terms*, first published in 1968.[2] This second edition contains over 900 terms, the vast majority of them referring to rhetorical figures. Its preparation included rechecking the existing entries in primary sources and in similar compilations published since 1967. After most of this work had been done by graduate students, over a period of several years, I reviewed their notes, rechecked pronunciations, etymologies, and cross-references, and revised numerous entries. From this work I gained a historical overview of how rhetorical figures and the names for them had proliferated down the centuries from Aristotle onward through Tudor England and into modern times.

But I do not propose to take you on a fatigue march through the history of defining and classifying the figures. Language is infinitely rich, and so are the possibilities for manipulating it artfully and cataloging the devices invented to do so. Nor will I talk about the theory of stylistic ornament—about whether there is any such thing as "natural language" that uses no figures, and figures are a

---

[1] *Alberici Casinensis flores rhetorici*, ed. D. M. Inguanez and H. M. Willard, Miscellanea Cassinese 14 (Montecassino, 1938). A new edition and a translation are being prepared by Thomas Coffey.

[2] Richard A. Lanham, *A Handlist of Rhetorical Terms* (Berkeley, 1968; 2nd ed. 1992).

garnish applied on top of pure meaning,[3] or whether there is an organic relation between ornament and sense, with figures functioning as channels or mirrors of thoughts and emotions.[4] Here, in the framework of this book's theme, is what interests me: After some 2,400 years of accumulated efforts to define and codify them, the underlying structure—the design system—of the rhetorical figures in Western literature is still deeply flawed.

HISTORY OF RHETORIC, DEFINITIONS OF TERMS

First, a quick glance at the early history of rhetoric and the place of the figures in it, and a few definitions of terms.[5] Rhetoric—most simply defined as the art of speaking well—was among the very first subjects of inquiry to be developed systematically: our oldest extant textbook is by Aristotle. Following Aristotle's successor Theophrastus (c. 370–c. 285 B.C.), rhetoric was divided into five parts according to function: *invention* (finding your arguments, figuring out what you want to say), *arrangement* (putting the parts of your speech in good order), *style* (choosing appropriate language, the best way to express each thought), *memory*, and *delivery*. The complete five-part system of rhetoric is most fully (and best) set forth in the twelve books of Quintilian's *Institutio oratoria* on the education of an orator, published about A.D. 95; this system formed the core of the curriculum in Western education until well into the nineteenth century.

By far the most highly developed parts of the system were invention and, above all, style, which includes rhetorical figures, the ornaments of style, as well as such matters as choice of words, sentence structure and rhythm, and kinds or levels of style. The emphasis on style soon led to the proliferation of figures and of handbooks cataloging them. Most of the Hellenistic Greek works on the subject are lost; the oldest preserved text devoted to style and the figures is found in book 4 of the Latin *Rhetorica ad Herennium* (after 85 B.C.; attributed to Cicero throughout the Middle Ages as *Rhetorica secunda* or *nova*), whose author thought this part of rhetoric so important (3.1.1) that he postponed discussion of it to his final book. By Quintilian's time, grammar teachers were teaching certain of the rhetorical figures at the primary-school level, and several Latin grammars, most famously the *Ars maior* of Donatus (mid-fourth century), include extended sections on figures.[6]

---

[3] See Brian Vickers, *In Defence of Rhetoric* (Oxford paperback, 1989), p. 227, quoting C. S. Baldwin, *Medieval Rhetoric and Poetic (to 1400)* (New York, 1928), pp. 181–82: "Writers as different as John of Salisbury and Brunetto Latini seem to think of [rhetoric] as polishing, decorating, especially dilating, what has been already expressed. It comes in after the real job is done; it has lost its ancient function of composing."

[4] Vickers, *Defence*, chapter 6, "The Expressive Function of Rhetorical Figures."

[5] George A. Kennedy's *Classical Rhetoric and Its Christian and Secular Tradition from Ancient to Modern Times* (Chapel Hill, 1980) provides a good general survey, with generous bibliography for further study.

[6] The oldest extant Latin grammar is that of Sacerdos, tentatively dated to the second half of the third century. It includes the faults and figures. (For biographical information about the Latin

In brief, classical and medieval Latin education is characterized by an intense focus on style. Today we can hardly imagine the attention paid to the choice and artful arrangement of words, to analyzing language in such minute detail. Gardens and flowers supplied a favorite metaphor for discussing stylistic ornament;[7] I have mentioned Alberic of Monte Cassino's *Flores rhetorici*, and a Tudor handbook was titled *The Garden of Eloquence*. We, having lost this intoxicating delight in words and their possibilities, tend to dismiss the garden as pure metaphor, but for some classical and medieval writers, the elements of stylistic ornament seem almost to take on an objective reality. As Geoffrey of Vinsauf says in the *Poetria nova* (1218 ff.), "The exercise given above has gathered together the flowers of diction. . . . In this way, then, let the mind's finger pluck its blooms in the field of rhetoric. But see that your style blossoms sparingly with such figures. . . . From varied flowers a sweeter fragrance rises . . ." and (1584 ff.), "Bring together figures of diction and thought, that the field of discourse may blossom with both sorts of flowers, for a mingled fragrance, blending adornment of both kinds, rises and spreads its sweetness."[8]

Like any proper garden, the garden of eloquence has a design, but it more resembles a casual, undisciplined, country garden than a formal one, and it is hedged round with thornbushes of definition. Its "flowers" are the many rhetorical devices broadly classified as *figures* or *schemes* (Latin *figura* translates Greek *schema*). Originally, both terms were very general, referring merely to the form or shape of anything;[9] *figura* was not used in a technical sense specific to language before Quintilian (the *Ad Herennium* uses *exornatio*). Quintilian, after acknowledging hot debate about the matter, defines *figura* as any form of language other than the usual or ordinary (9.1.4, 11), or any form of expression (*forma dicendi*) somehow made new by art (9.1.14).

Most often, the figures are then subdivided (e.g. Quintilian 9.1.17) into:

• *figures of speech* (or of words, diction, expression, or language; in Greek, *schemata lexeos*. "It is a figure of diction if the adornment is comprised in the fine polish of the language itself," *Rhetorica ad Herennium* 4.12.18);

• *figures of thought* (or of the mind, feeling, or conceptions; Greek *schemata dianoeas*. "A figure of thought derives a certain distinction from the idea, not from the words," *Rhetorica ad Herennium* 4.12.18);

• *metaplasms* (change in the form [i.e. spelling] of a word); and

• *tropes* (change in the meaning of a word: "the artistic alteration of a word or phrase from its proper meaning to another," Quintilian 8.6.1; "a figure that

---

grammarians, see Robert A. Kaster, *Guardians of Language: The Grammarian and Society in Late Antiquity* [Berkeley, 1988].)

7 See, e.g., La Rue Van Hook, *The Metaphorical Terminology of Greek Rhetoric and Literary Criticism* (Chicago, 1905), pp. 17–18.

8 *Poetria Nova of Geoffrey of Vinsauf*, trans. Margaret F. Nims (Toronto, 1967), pp. 60, 72.

9 Quintilian states that the term *figura* has two senses: "In the first it is applied to any form in which thought is expressed, just as it is to bodies which, whatever their composition, must have some shape. . . ." (9.1.10). Translations of Quintilian are from the Loeb Library edition by H. E. Butler.

changes the meaning of a word or words, rather than simply arranging them in a pattern of some sort," Lanham, *Handlist*, s.v.).

All four categories are usually organized (not always explicitly) according to four possible kinds of change: addition, subtraction, mutation, and transposition. Some writers make absurd and untenable distinctions. A figure may be treated as a figure of diction by one writer but as a figure of thought by another. Figures of speech may be subdivided into forms of words (orthographical) and arrangement of words (syntactical); Quintilian says the former category is more grammatical, the latter more rhetorical (9.3.2). They may concern individual words or groups of words. Tropes are sometimes classed as a subdivision of figures, sometimes treated as a separate category altogether. And, however all these categories are set up, they are contrasted with *faults*, with *vitia*.

As all gardeners know, gardens do tend to have weeds among the flowers. Weeds offer a useful metaphor for making sense of the riot of figures, for weeds and figures share certain characteristics. Consider:

• The flowers of weeds are often pretty but harmful. / Figures have a dangerous allure in their power to move the emotions.

• How does one decide that the pretty flower is a weed and not a cultivar? / When does a figure become a fault? When does a figure of words become a figure of thought?

• Weeds spread and drive out desired plants. / An overabundance of figures overwhelms and ruins an artful design.

• Weeds are good at surviving and multiplying—and so were the figures.

From the beginning, the impulse to invent new terms was simply irresistible. Already by the end of the first century A.D., Quintilian could express exasperation with the unruly proliferation of rhetorical figures: "Writers have given special names to all the figures, but variously and as it pleased them" (9.3.54; see also 9.1.22–24). In Elizabethan England, fifteen hundred years later, nothing had changed: Tudor rhetoricians published handbooks left and right, in which they retained Greek and Latin names for the figures. (One maverick, George Puttenham, substituted colorful English replacements such as "Fleering Frumpe," "Cuckowspell," and "Drie Mock," but unfortunately—or perhaps fortunately!—they didn't catch on.) In fact, the Greek and Latin terms have never been seriously challenged, and they are still in general use today.

Before I suggest some reasons for this proliferation of figures, I will describe their typical production formats. Finally, I will examine what seems to me the system's most interesting design flaw.

PRODUCTION FORMATS

Although Aristotle discusses several aspects of style and a few figures in book 3 of his *Rhetoric*, his cannot be called a systematic treatment, and he has no general term for "figure of speech." (He is especially interested in metaphor, and also mentions a few of what would become known as the Gorgianic figures:

antithesis, balance, and assonance.)[10] As I have already noted, very little remains of works on rhetoric produced between Aristotle's time and the *Ad Herennium* of the early first century B.C., but it is clear that the *Ad Herennium* depends heavily on earlier Greek works on style and figures, probably including free-standing compendia of figures such as its own book 4.

The typical format for a work devoted to the figures is a catalog, either self-contained (e.g. the anonymous *Schemata dianoeas*, Bede's *De schematibus et tropis*) or embedded in a comprehensive rhetoric textbook (*Rhetorica ad Herennium*, Quintilian) or, later, in a grammar (Donatus and his successors). An occasional treatise is in verse, and Onulf of Speyer (fl. c. 1050) produced both prose and verse versions of his *Colores rhetorici* (thankfully, he uses the same terms in both). One or more examples are given for each figure defined, most of them being taken from classical poets, above all Vergil; of prose authors, only Cicero ever furnishes an appreciable number of examples. The order of presentation is not fixed; the *Ad Herennium* author covers figures of diction first, then figures of thought, but Quintilian reverses the order, on the grounds that "in the natural course of things we conceive ideas before we express them" (9.1.19). Figures are also pointed out, but not systematically, in commentaries on literary works such as those by Servius and Donatus on Vergil.

Modern composition handbooks strive mightily to differentiate themselves one from another and to hold the student's interest by concocting interesting or amusing examples.[11] Medieval rhetoric handbooks, on the other hand, contain a vast amount of what we might call plagiarism: the same examples, and often the same definitions, are used repeatedly by successive generations of writers, even when they change the names of the terms. It seems likely that the tendency to copy was present from the Greek beginnings. Historians of rhetoric have often noted that Christian writers on the figures were very slow to substitute examples taken from Scripture, not to mention the growing body of Christian literature.[12] If Donatus was a Christian you would never know it from his grammar. Only in the sixth century did Cassiodorus show how to teach rhetoric from the Bible, by regularly pointing out rhetorical figures in his large commentary on the Psalms,[13] and Bede was the first to systematically use examples from the Bible in a

---

[10] See George Kennedy, *The Art of Persuasion in Greece* (Princeton, 1963), pp. 103–14; and his translation of the *Rhetoric* (New York, 1991), p. 242.

[11] An extreme example is Karen Elizabeth Gordon, *The Deluxe Transitive Vampire: The Ultimate Handbook of Grammar for the Innocent, the Eager, and the Doomed* (New York, 1993), featuring a cast of bizarre characters and impudent, even risqué, examples that make the conventional rules memorable.

[12] The Christian attitude toward classical rhetoric has of course been much studied. A convenient starting point is M. L. W. Laistner, *Thought and Letters in Western Europe, A.D. 500 to 900*, 2nd ed. (Ithaca, 1966), Index s.v. Rhetoric. For more detail, see James J. Murphy, *Rhetoric in the Middle Ages* (Berkeley, 1974), especially chapter 2, "Saint Augustine and the Age of Transition, A.D. 400 to 1050."

[13] See James J. O'Donnell, *Cassiodorus* (Berkeley, 1979), pp. 159–62; for the dating (in the 540s), pp. 134–36 and 168–72.

textbook on the figures.[14] Most later writers continued to recycle examples from classical literature, however.

In fact, the entire medieval tradition of the figures rests on three pagan texts: book 4 of the *Rhetorica ad Herennium*, books 8 and 9 of Quintilian, and, above all, book 3 of Donatus's grammar. Throughout the early Middle Ages, while more advanced rhetoric was hibernating, the classical figures continued to be studied with relish as part of the grammar school curriculum. We can see something of that curriculum in several large compendia manuscripts, of the eighth and ninth centuries, that contain Donatus and other grammars, commentaries on Donatus, and discrete collections of rhetorical figures.[15] One of the earliest such manuscripts is Paris, Bibliothèque Nationale, MS lat. 7530, from the end of the eighth century, the first datable manuscript in the Beneventan hand; it has been called an encyclopedia of the liberal arts.[16]

THE PROLIFERATION OF TERMS

What encouraged the rhetorical figures to thrive and multiply? I see four chief causes.

(1) The first surely is familiar to all academics: *the scholar's zeal to define, refine, and subdivide.* Quintilian warned against the indiscriminate naming of figures (e.g. 9.3.99), but in vain. The manifestations of this impulse take various forms.

(a) Quintilian refers to "the various names which the Greeks . . . are so fond of inventing" (9.1.22), and Hellenistic Greek scholars were probably responsible for many of the almost synonymous terms and razor-fine distinctions we find among the Greek terms.

> asyndeton, dialyton
> epiphora, epistrophe
> homoioteleuton, homoioptoton

In the cluster of ten terms assembled under *Macrologia* in Lanham's *Handlist of Rhetorical Terms*, all but one are Greek in form, and in practice it is often very difficult to decide which of them best fits the case in hand.

(b) Latin rhetorical theory was in turn modeled on Greek. Early Latin writers, rather than borrow existing Greek terms, tried to invent Latin terms

---

[14] In his *De schematibus et tropis*, ed. Karl Halm, *Rhetores latini minores [RLM]* (Leipzig, 1863). "[H]is principle of co-ordinating the rhetorical theory of figures with the study of the Bible prevailed and was to grow like a mustard-seed," E. R. Curtius, *European Literature and the Latin Middle Ages*, trans. Willard R. Trask, Bollingen Series 36 (New York, 1953), p. 47.

[15] Full descriptions are given by Louis Holtz in his monumental edition of Donatus, *Donat et la tradition de l'enseignement grammatical: Etude sur l'"Ars Donati" et sa diffusion (IVe–IXe siècle) et édition critique* (Paris, 1981).

[16] Louis Holtz, "Le Parisinus Latinus 7530, synthèse cassinienne des arts libéraux," *Studi medievali* 3rd ser. 16 (1975), 97–152 at 99.

for the figures, and so almost every Greek term eventually acquired at least one Latin doublet.[17]

> epanaphora / repetitio, relatio
> hyperbaton / transgressio
> periphrasis / circu(m)itio

Usually the Greek term remained dominant; *simile* is a rare case of Latin's winning out, over Greek *eicon*.

(c) Both languages were rich in prefixes, and so doublets differing only by prefix appear in both languages.

> Greek: ecphonesis, epecphonesis, epiphonesis
>   prothesis, prosthesis
>   synzeugmenon, epezeugmenon
> Latin: dinumeratio, enumeratio
>   obticentia, reticentia

(d) Some Latin doublets and near-doublets may stem from uncertainty or disagreement about the correct or best translation of a Greek term.

> Aposiopesis, quam idem Cicero reticentiam, Celsus obticentiam, nonnulli interruptionem appellant . . . (Quintilian 9.2.54)

> synathroismos, symphoresis / frequentatio (*Ad Her.* 4.52), conductio (Cicero, *De inv.* 1.74), congeries (Quintilian 8.4.26), consummatio (Quintilian 9.2.103)

The domain of style offers other examples of the urge to subdivide and catalog: 5 or 21 or 45 kinds of letters, 21 types of sententiae, Hermogenes' 20 types of style; the meters of classical poetry. The itch to name and classify is a way of "deconstructing" the literary work, of tearing away its veil of artifice (this is another metaphor widely used of rhetoric—as cosmetic or harlot) and thus acquiring a sense of intimacy with, and power over, the work.

(2) *Errors by scribes, early printers, and modern scholars* compose a second source of proliferation of terms. Some, I suspect, arise from that scholarly zeal just outlined.

(a) Medieval scribes often mistranscribed Greek letters. Abundant examples that suggest how the corruption proceeded can be found in the apparatus criticus of any modern edition that tries to replicate scribal letter forms. Figure 1 reproduces a few lines of the text and apparatus criticus,

---

[17] The *Ad Herennium* uses *no* Greek terms, Cicero very few: Henri Bornecque, "La Façon de désigner les figures de rhétorique dans la *Rhétorique à Hérennius* et dans les ouvrages de rhétorique de Cicéron," *Revue de philologie*, 3rd ser., 8 (1934), 141–58 at 158. In his Loeb Library edition of the *Ad Herennium*, Harry Caplan supplies the Greek equivalents in the notes to book 4.

containing three Greek terms, of Michael Winterbottom's 1970 Oxford edition of Quintilian; I shall discuss below the rejected ninth-century reading *soraismos*, which survived into Elizabethan handbooks.

Sunt inornata et haec: quod male dispositum est, id ἀνοικονόμητον, quod male figuratum, id ἀσχημάτιστον, quod male conlocatum, id κακοσύνθετον uocant. Sed de dispositione diximus, de figuris et compositione dicemus. Σαρδισμός quoque appellatur quaedam mixta ex uaria ratione linguarum oratio, ut si Atticis Dorica et Aeolica et Iadica confundas. Cui simile uitium est apud nos si quis sublimia humilibus, uetera nouis, poetica uulgaribus misceat— ...

21 anoiKON OMHTON *G*: ΑΠΟΙΝΟΜΗΤΟΝ *A*: ΑΝΟΕCΟΜΕΤΟΝ *Π*   ἀσχημάτιστον *1418, ut coni. Halm (cf. 9. 1. 13)*: ΑCΧΗΜΑΤΟΝ *A*: scheamastiton *Π*   23 σαρδισμός *Halm ex Π*: COΡΑΙCΜΟC *A*   24 quoque: ⟨sic⟩ *Kiderlin 1891-2*   25 Dorica et *scripsi*: loricae *G* (ionica et *a in ras.*): dorica *Π*   aeolica et *iadica A* (aeolica etiam dicta *G*): eolica iadicas *Π*

Fig. 1. Quintilian, *Institutio oratoria* 8.3.59–60 (ed. Michael Winterbottom, Oxford, 1970): text, with apparatus criticus reporting eighth- and ninth-century manuscript renderings of Greek terms.

(b) Early printers occasionally created new rhetorical terms by introducing errors when printing Greek letters.[18] Usually these were short-lived, being recognized as errors, but such errors could be perpetuated if misprints were handed on from one manual to another. In place of the correct form *antistoechon*, for example, Richard Sherry's *Treatise of Schemes and Tropes* (1550, fol. B6v), by changing only one letter, prints the impossible form *antisthecon* (no such root exists in Greek). Probably the printer, rather than Sherry himself, was responsible.[19]

---

[18] La Rue Van Hook, "Greek Rhetorical Terminology in Puttenham's *The Arte of English Poesie*," *TAPA* 45 (1914), 111–28 at 118: "Numerous errors in the transliteration of the terminology are to be observed in Book III of Puttenham's treatise. . . . Some of these errors are obviously printer's mistakes, for the Greek terms seem to have puzzled the typesetter" (he lists them).

[19] I cannot prove this assertion, of course, without knowing exactly what printings of what source-texts Sherry used. A possible source for *antisthecon* is Petrus Mosellanus's *Tabulae de schematibus* (1536)—Sherry mentions Mosellanus as a source in his dedicatory epistle, fol. A5r—but I have not been able to see a copy of Mosellanus. Another possible source, not named by Sherry, is a popular handbook first published in 1541 by Joannes Susenbrotus, *Epitome troporum ac schematum et grammaticorum et rhetoricorum*. Early copies are extremely rare; in the undated

(c) Modern scholars who are too respectful of their source texts may perpetuate existing misprints, or even introduce new errors. Sherry's impossible *antisthecon* was picked up by two twentieth-century students of rhetorical terms, Warren Taylor and Sister Miriam Joseph.[20] Sherry's 1550 *Treatise* contains another error, *oictros* for *oictos* (where the intrusive *r* came from, I do not know), which Taylor repeated and compounded by printing "Cictros," misreading as *C* the initial capital letter *O* of the italic font used by the printer for the technical terms (figure 2).[21]

(1) Another forme is called *Oictros*, oꝛ cōmiſeracion, wherby teares be

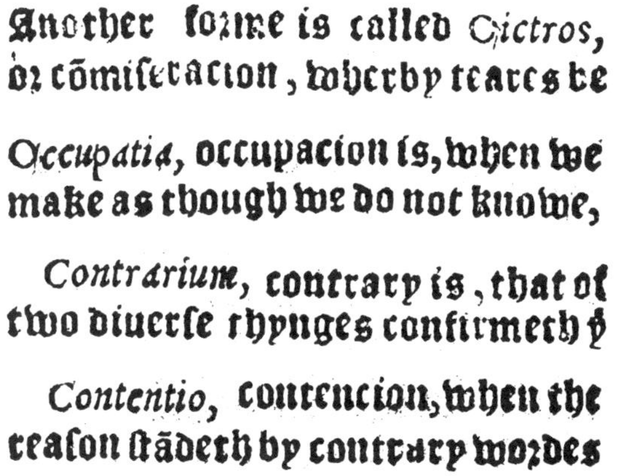

(2) *Occupatia*, occupacion is, when we make as though we do not knowe,

Contrarium, contrary is, that of two diuerſe rhynges confirmeth ꝑ

Contentio, contencion, when the reaſon ſtādeth by contrary woꝛdes

Fig. 2. From Richard Sherry, *A Treatise of Schemes and Tropes* (1550): (1) *Oictros*, transcribed by Taylor, *Tudor Figures of Rhetoric* (1937) as *Cictros*; (2) samples of italic capital *O* and *C* from elsewhere in Sherry's text. Note printer's error *occupatia* for *occupatio*. Photo: The Huntington Library.

(3) *"Turf wars" between grammar teachers and rhetoric teachers* in the late antique period probably generated some doublets for existing rhetorical terms, perhaps even some new terms. Theoretically, in Latin schools of the Augustan

---

edition I have seen, fol. B5r, Susenbrotus quotes lavishly from the *Carmen de figuris* (written in dactylic hexameters) by Antonio Mancinelli (d. 1505), and under the heading *Antithesis* both give the correct form *antistoechon* (as a "more suitable," *aptius, apte,* term for the figure in question— illustrating once again the phenomenon of doublets).

For the career and textbooks of Antonio Mancinelli, see Paul F. Grendler, *Schooling in Renaissance Italy: Literacy and Learning, 1300–1600* (Baltimore, 1989), pp. 183–86; Richard Sherry, *A Treatise of Schemes and Tropes* (1550), facsimile repro., intro. Herbert W. Hildebrandt (Gainesville, Fla., 1961).

[20] Warren Taylor, *Tudor Figures of Rhetoric* (1937; repr. Whitewater, Wisc., 1972), p. 71; Sister Miriam Joseph, *Shakespeare's Use of the Arts of Language* (New York, 1947), pp. 53, 294.

[21] Sherry, fol. E2v; Taylor, *Tudor Figures*, p. 82.

era, the *grammaticus* handled figures of words, the "grammatical" figures, and tropes in the course of explicating poetry, and then the rhetoric teacher taught figures of thought (the orator, of course, must command both categories). This division of labor between grammaticus and rhetor is still given lip service in later Latin grammars such as Donatus.[22] But already in Quintilian's time, grammar teachers were taking over the first exercises in rhetoric, and so tended to take over the figures of thought as well.[23] When he is discussing early education in grammar, Quintilian merely states that the grammatici are to teach the whole works—metaplasms, the schemata lexeos *and dianoeas*, vitia, and tropes—but that he will postpone discussing *all* of this until he talks about the rhetoric curriculum, *de ornatu orationis* (1.8.14–16). But, unlike the *Ad Herennium* author, he gives no reason for the postponement, and indeed he subsequently notes that the two groups of teachers disagree about various terms and definitions (e.g. 8.5.35–6.1). This little turf war signals something large: the late antique decline of advanced rhetoric studies and the rise of the medieval "grammar school" curriculum. Isidore's *Etymologies* reflects the blurring of the division by including sections on rhetorical figures in both book 1 (grammar) and book 2 (rhetoric).

(4) Finally, *many figures or "virtues" are also identified, under different names, as "*vices," *vitia*. How can this be, that a figure, by definition an *ornament* of style, may be considered a fault? Here we surprise a fundamental contradiction at the heart of *ornatus*—one that has never been resolved. It is a fairly old question, having bedeviled the subject at least since Pliny the Elder, who died in A.D. 79. If figures are flowers in the garden of eloquence, faults surely qualify as its weeds. The metaphor of weeds is my own, but in discussing this flaw in the design system of rhetorical figures in the remainder of this essay, I am much indebted to a brilliant analysis by Marc Baratin and Françoise Desbordes.[24]

FIGURES AND FAULTS, WEEDS AND FLOWERS

The grammaticus, the Latin grammar school teacher, had two basic responsibilities: to teach the principles of speaking and writing Latin correctly,

---

[22] See, for example, Aquila (3rd c. A.D.), *De figuris sententiarum et elocutionis*, ed. Halm, *RLM*, p. 22.11–12: "Figurandarum sententiarum et elocutionum [= dianoeas, lexeos] proprium oratoris munus est"; Donatus, *Ars maior* 3.5: "schemata dianoeas ad oratores [Julian of Toledo, *De vitiis et figuris*, adds, "id est ad philosophos"] pertinent, ad grammaticos lexeos"; Servius, *in Don.*, GL 4.448.1–2: "Plane sciendum est quoniam schema in sermone factum ad grammaticos pertinet, in sensu factum ad oratores."

[23] Louis Holtz analyzed this rivalry with great clarity in an important article that bears on the "structural design flaw" I am trying to describe: "Grammairiens et rhéteurs romains en concurrence pour l'enseignement des figures de rhétorique," *Colloque sur la rhétorique: Calliope I*, ed. R. Chevallier (Paris, 1979), pp. 207–20.

[24] Marc Baratin and Françoise Desbordes, "La 'troisième partie' de l'*ars grammatica*," *Historiographia linguistica* 13.2/3 (1986), 215–40.

and to explicate the poets, chiefly Vergil.[25] Grammar was founded on a theory of correctness, *hellenismos* or *latinitas*, thought to have been developed by Stoic grammarians. Correctness naturally included its opposite, faults or vitia. Discussions of the vitia always begin with *barbarism* and *solecism* (both, Greek terms carried over to Latin). A barbarism is a fault in a single word, such as the omission of a letter or syllable; a solecism is an error involving two or more words, such as a disagreement between subject and verb. Along with barbarism and solecism, Latin grammars identify and define ten or so other faults, such as ellipsis, pleonasm, and amphibologia (ambiguity).

Ah, but *poets* do such things all the time—think of Vergil, author of the national epic and number one source for examples of the figures! Since the chief material for instruction was poetry, the grammaticus immediately had to explain away the licenses allowed to verse—to confront the problem that poets used words, forms, and constructions different from those of everyday speech. Correctness was thus undermined—vitiated!—by *ornatus*; the study of literary style thus began with *faults*, and the paradox of having continually to hedge: It's all right for Vergil to say this, my boy, but you must not imitate it in your compositions! How, then, was the poor child to learn what *was* permissible and desirable? Where were standards to be found?

Servius reports, in his commentary on Donatus's grammar, that the question was raised with Pliny the Elder:

> What is the difference between figures and faults? For although figures are used for ornament, and faults are avoided, and yet the same examples are found for both figures and faults, there must be some difference. Therefore [runs the answer], whatever we do intentionally *(scientes)*, in search of the new or unusual, and which is supported by examples from appropriate authors, is called a figure. But whatever we set down out of ignorance is considered a fault. . . . [26]

An example from the *Aeneid* follows, of a singular subject with a plural verb. "If someone says this ('pars in frusta secant') intentionally *(sciens)* and for the sake of variety, he creates a figure; but if out of ignorance he joins numbers together

---

[25] See, e.g., Quintilian 1.4.2: "Haec igitur professio [of the *grammaticus*], cum brevissime in duas partes dividatur, recte loquendi scientiam et poetarum enarrationem . . ."

[26] Servius, *in Don.*, GL 4.447.5–10 (quoted by Baratin and Desbordes, pp. 237–38; my translation): "Quaesitum est apud Plinium Secundum, quid interesset inter figuras et vitia. nam cum figurae ad ornatum adhibeantur, vitia vitentur, eadem autem inveniantur exempla tam in figuris quam in vitiis, debet aliqua esse discretio. quidquid ergo scientes facimus novitatis cupidi, quod tamen idoneorum auctorum firmatur exemplis, figura dicitur. quidquid autem ignorantes ponimus, vitium putatur." The words "apud Plinium" suggest that someone other than Pliny himself raised the question. (Perhaps Pliny addressed this problem in one of two lost works named by his nephew [Epist. 3.5], *Dubii sermonis octo* or *Studiosi tres*, the latter "in sex volumina propter amplitudinem divisi, quibus oratorem ab incunabulis instituit et perficit" [divided into six rolls on account of its size, in which he trains the orator from infancy to perfect mastery]—a model for Quintilian's *Institutio*, or a challenge?)

incongruously [i.e. singular and plural, as in the example Servius has just given] when he means something different, he is judged to have committed a solecism."[27]

Attempting to distinguish figure from fault by the writer's *intent* may seem to you both impossible and foolish, but because Quintilian, Pliny's contemporary, repeatedly fell into the same trap, both of them may be parroting a current theory. Quintilian says, for example, that figures of speech "originate from the same sources as errors of language. For every figure of this kind would be an error, *if it were accidental and not deliberate*" (my emphasis).[28] Later grammarians adopt the same explanation.[29]

Elsewhere, in his commentary on the *Aeneid*, Servius refines Pliny's distinction into a *three*-part taxonomy. Again it is a matter of a singular subject with a plural verb. Here, after defining metaplasm as a deliberate fault *(ratione vitiosus)* in one word, and figure as a deliberate fault in connected words *(contextu sermonum)*, he adds that metaplasm and figure "are *in the middle* [between correctness and fault], and they are differentiated by skill [or, knowledge] and ignorance *(peritia et inperitia)*."[30]

A much clearer account of this three-part taxonomy is given by John of Salisbury in his *Metalogicon*, when he discusses grammar and the grammarian's duties (1.17–18). Grammar itself has three parts: the grammatical art (i.e. the rules of correct speech), grammatical errors, and, "because poetry belongs to grammar," figures:

> The figure [of speech], however, occupies an intermediate position.
> Since it differs to some extent from both [regular grammar and

---

[27] Ibid., 10–13: "Si sciens quis dicat 'pars in frusta secant' et causa varietatis hoc dicat, figuram facit; si autem nescius, cum aliud velit dicere, incongrue inter se numeros iunxerit, soloecismum fecisse iudicatur."

[28] "Esset enim omne eiusmodi schema vitium si non peteretur sed accideret" (9.3.2). See also 8.3.50–51, "Sed hoc *[elleipsis]* quoque, *cum a prudentibus fit*, schema dici solet," and 1.5.53, "hic quoque quod schema vocatur, *si ab aliquo per inprudentiam factum erit*, soloecismi vitio non carebit."

[29] E.g. (claimed for) Victorinus, frag. *De soloecismo et barbarismo,* ed. Maximilian Niedermann as addendum to his edition of Consentius, *Ars de barbarismis et metaplasmis* (Neuchâtel, 1937), p. 35.16–18: "Numquam ergo soloecismus excusari potest. si a nobis *per inprudentiam* fiat, vitium est; si a poetis vel oratoribus *affectate* dicatur, figura locutionis et appellatur graece *schema*"; repeated at p. 37.3–5 for barbarismus (omitting *affectate* and substituting *metaplasmus*). Audax is identical for barbarism (GL 7.362.19–21), but his text for solecism is missing. Isidore, *Etym.* 1.35.7.

[30] Servius, *in Aen.* 5.120 (quoted in this connection by Holtz, *Donat* [see above, n. 15], pp. 148–49): "PUBES INPELLUNT figura est, ut [1.212] *pars in frusta secant.* et sciendum inter barbarismum et lexin [Isidore, *Etym.* 1.35.7, substitutes *figuras*], hoc est, Latinam et perfectam elocutionem, metaplasmum esse, qui in uno sermone fit ratione* vitiosus. item inter soloecismum et schema, id est, perfectam sermonum conexionem, figura est quae fit contextu sermonum ratione* vitiosa. ergo metaplasmus [*-mi*, Isid.] et figura [*schemata*, Isid.] media sunt, et discernuntur peritia et inperitia. fiunt autem ad ornatum." *At *Etym.* 1.35.7, Lindsay inexplicably prints *oratione* both times, though he reports *ratione* from three manuscripts.

grammatical error], it falls in neither category. All strive to conform to the [grammatical] art, since it is commanded, and to shun [grammatical] mistakes, since these are forbidden; but only some use figures, since the latter are [merely] permissible. Between errors, that is to say, barbarisms and solecisms, and the art [of grammar], which consists in normal good speech, stand figures and schemata. . . . License to use figures is reserved for authors and for those like them, namely, the very learned.[31]

In other words, figures are neither right (commanded) nor wrong (forbidden), but optional (permissible). Permissible, that is, for those who are "very learned," *scientes*. "Natural" speech is plain and unornamented—the untutored cannot possibly be eloquent. This is a very dead end.

A more common, but equally witless, approach is represented by Donatus, grammarian to the Middle Ages. Book 3 of his grammar (known as *Barbarismus*) treats faults and figures in six chapters that divide neatly (too neatly, as Baratin and Desbordes prove) into two halves: barbarism, solecism, and other faults; metaplasm, schemes, and tropes. Donatus defines barbarism and solecism thus: "A barbarism is one part of speech [i.e. one word] that is faulty *in communi sermone* [which normally means prose]. In poetry [it is called] a metaplasm" (3.1); his definition of solecism ends, "A solecism in prose *[in prosa oratione]* is called a scheme in poetry" (3.2). Presto! Terminology cures the split, by making each contrasted pair—barbarism and metaplasm, solecism and scheme—interchangeable, and *all four* categories deviations from the standard of correctness.[32]

This might work—disregarding for the moment the distinction between poetry and prose—if only Donatus (and others) did not use *exactly the same examples* to illustrate figure and fault. I'll give just one illustration, from Donatus:

> Fault (barbarism): addition of letter or syllable, e.g. *relliquias Danaum*
> Figure *(epenthesis)*: addition of letter or syllable, e.g. *relliquias*[33]

Two things to note here. First, this example of barbarism by addition of a letter, like the two other examples he gives, comes from poetry, and the *Aeneid* (1.30) at that. Must we not infer that Donatus is accusing Vergil of committing barbarisms? Second, in the section on figures a few pages later, this same phenomenon—addition of a letter or syllable—has three subdivisions, depending on where in the word it occurs, and for all three he records alternate names, doublets:

---

[31] *The Metalogicon of John of Salisbury*, trans. Daniel D. McGarry (Berkeley, 1962), p. 54. The bracketed words are McGarry's supplements.

[32] Holtz, *Donat*, pp. 148–49.

[33] Ibid.; full text at pp. 653.8 and 661.1–2.

(1) addition at the beginning of a word is *prothesis* or *prosthesis* (e.g. *gnato*)

(2) in the middle, it is *epenthesis* or *parenthesis* (*relliquias* again)

(3) at the end, *paragoge* or *prosparalempsis* (the example is *potestur*)

As this example suggests, analysis of the figures, and hence their terminology, are much more fully developed than those of the faults. (The development of a canon of faults deserves a separate study.)

Consentius (after Donatus; date uncertain) seems to have been the only grammarian to recognize and protest the confusion induced by using the same examples to illustrate both figure and fault.

> I shall not imitate those writers who have chosen to give examples of this kind of faults [i.e. barbarisms] from literature *(de auctoritate lectionum)*. The result has been such confusion that almost no one now understands what is a barbarism and what is a metaplasm. For often some call it a metaplasm while others call it a barbarism, and sometimes the same people, using the very same citation, call it both a metaplasm and a barbarism, and so they confound everything.[34]

Instead, he will illustrate the faults from the spoken language of everyday speech: "I shall give examples of [barbarism] which we can notice in everyday speech if we listen a little more carefully "[35]

One last line of approach remains: If it occurs in poetry, no problem—it's a metaplasm or a figure; if in prose, it's a fault—a barbarism or a solecism.[36] This serves only to confirm that grammar instruction in the classical period paid little attention to analyzing prose authors—even though elementary students wrote prose compositions and the ultimate goal of rhetorical education was to produce an orator who would deliver speeches exclusively in prose. What little formal study of prose did take place in school was the rhetoric teacher's responsibility.

One always wants to see what Quintilian has to say about a problem, because he is certainly intelligent and usually sensible. Not so with the oneness of fault and figure. Ellipsis, he says, the omission of words, "may be either a blemish or a figure, according to the context" (9.3.18). The repetition of a word or phrase, when regarded as a fault, is called *tautologia*; when the same phenomenon is ranked among the figures, it is called *epanalepsis* (8.3.51). *Pleonasm*, as in "I

---

[34] "Non imitabor eos scriptores qui exempla huius modi vitiorum de auctoritate lectionum dare voluerunt; quo ea vitiorum facta est confusio ut paene iam nemo intelligat quid barbarismus sit, quid metaplasmus. Nam plerumque alii atque alii, interdum iidem ipsi, et metaplasmum et barbarismum dicentes eiusdem lectionis utuntur exemplis, eoque cuncta confundunt." Quoted from GL 5.391.26–31 by Baratin and Desbordes, pp. 222–23.

[35] "Nos exempla huius modi dabimus quae in usu cotidie loquentium animadvertere possumus si paulo ea curiosius audiamus," GL 5.391.31–33. (His examples provide rare and valuable evidence for the pronunciation of Latin in northern Africa in the late fourth or early fifth century.)

[36] E.g., Servius, *in Don.*, GL 4.444.8–10: "praeterea si in prosa oratione fiat, tunc barbarismus dicitur; si autem in poemate, metaplasmus vocatur" (repeated for solecism, 447.2–4).

saw it with my own eyes," is a fault—but sometimes it has "a pleasing effect when employed for the sake of emphasis," as in "I heard his voice with my own ears" (8.3.54 and 9.3.46–47). Again and again, Quintilian fails to confront the problem. Or rather, he dismisses it, as when he declares, "Speech is corrupted in just as many ways as it is ornamented" (8.3.58, "Totidem autem generibus corrumpitur oratio quot ornatur"). Are we not with Alice and Humpty Dumpty in Wonderland, where everything means just what Humpty Dumpty chooses it to mean?

Cassiodorus is the only writer I have come across, besides Consentius, who was bothered by all these failures to cope with the question put to Pliny the Elder. In the *Institutions* (2.1.2) he objects to labeling as *faults* constructions devised for the purpose of ornament and validated by the best pagan authors *(auctores)* and especially, *maxime*, by the authority of holy Scripture, *divinae legis*. In his commentary on the Psalms he dispatches the problem—for the Psalter, at least— by implicitly denying that the words of Scripture *can* contain faults. He does this simply by ignoring constructions that could be considered faults, or by redefining them positively. *Verba divina* are by definition *perfecta*.[37]

I shall just mention another important area of rhetorical theory in which the problem recurs: that of the kinds or levels of style, *genera orationum*. A division into three appears first in the *Ad Herennium*: grand or high, plain or low, and middle, splitting the difference.[38] After characterizing each, the *Ad Herennium* author warns against falling into their corresponding *bad* styles (4.10.15). "Bad" in this case is easy to define: excessive, too much of a good thing. Brevity characterizes the plain style, for instance, but an excess of brevity leads to obscurity. This kind of excess has a name, naturally—*cacozelia*—and Quintilian declares it the very worst kind of offence against style, because it is done deliberately (8.3.56). But wait a minute: "done deliberately"—where have we heard that before? We are back in the starting box with Pliny the Elder! And indeed, the same kinds of argument and confusion mark the long debate about kinds of style.

One rhetorical term for going around in a circle—that is, for bringing a complex utterance to a conclusion— is "period" (Cicero used eleven others as well!).[39] By way of concluding, I want to exhibit an unusual but little-noticed weed.

I said that the weeds in the garden of eloquence include mutants. A most curious mutant appears among the faults catalogued by Quintilian at the beginning of his discussion of *ornatus* in book 8. He defines the term as the indiscriminate mingling of Greek dialects, and then compares it to the fault in Latin of mingling levels of diction by mixing lofty words with humble ones, old with new, and poetic with common. In some manuscripts and early editions of Quintilian (including the Aldine, 1512) and in Elizabethan handbooks this term

37 J.-M. Courtès, "Figures et tropes dans le Psautier de Cassiodore," *Revue des études latines* 42 (1964), 361–75 at 368–69.

38 See Caplan's note in the Loeb Library edition, p. 252.

39 Bornecque (see above, n. 18), p. 141.

appears as *soraismus*. This is the form transmitted by one of the oldest extant manuscripts of Quintilian; its reading is reproduced in the apparatus criticus in figure 1 above (Milan, Biblioteca Ambrosiana, MS E. 153 sup.; *siglum* A). The correct reading, however, appears to be *sardismus*—which is not transmitted by any extant manuscript of Quintilian. But *sardismus* is found in a yet older manuscript, BN lat. 7530 (*siglum* Π in figure 1), the late eighth-century encyclopedic compendium that I described earlier, in an anonymous collection of rhetorical figures called *Schemata dianoeas* that was cobbled together from three sources: Quintilian's two books on figures, Isidore's *Etymologies*, and an unidentified source. The latest editor of the collection believes that the Quintilian excerpts originated no later than the early sixth century,[40] and indeed the reading *sardismus* is confirmed by Cassiodorus three times in his commentary on the Psalms—where, in accordance with his principle that Scripture contains no faults, he makes *sardismus* not a fault but a figure: the (intentional) mixing of languages, such as Hebrew, Greek, and Latin. The term remained a virtue for medieval writers who used Cassiodorus,[41] but reverted to being a fault with the Elizabethans who drew directly on Quintilian (Puttenham called it "Mingle-Mangle").

It is easy to see how *sardismus* could be misread as *soraismus*, especially if, as is likely, it was originally written in Greek majuscule letters. I do not know the full textual history of Quintilian's *Institutio oratoria*, but it looks as though the Tudor rhetoricians received the form *soraismus* via the Aldine edition, which must have used a manuscript with that reading. Perhaps none of this would be of much interest, except that (1) *both* forms of the word are unique—neither *sardismus* nor *soraismus* has any congeners in Greek[42]—and (2) the vulgate reading in printed editions of Quintilian's text until Karl Halm's 1868 edition was a totally different form, *koinismus*. That too is a unique form, unattested elsewhere in Greek; I suspect that it was a Renaissance emendation, prompted by the revival of interest in Greek. Nowadays, however, *koinismus* does not even appear in the textual apparatus as a plausible but rejected reading—modern editors have uprooted it, eradicated it, cast it out again from the garden of eloquence.

UCLA CENTER FOR MEDIEVAL AND RENAISSANCE STUDIES

---

[40] Ulrich Schindel, ed., *Anonymus Ecksteinii, Scemata dianoeas quae ad rhetores pertinent*, Nachrichten der Akademie der Wissenschaften Göttingen, Phil.-hist. Kl. 1987 no. 7 (Göttingen, 1987), pp. 141–42.

[41] Chiefly Eugraphius (mid-sixth century?) or a later writer who made additions to his commentary on Terence, *Andria* 5.4.16 (*Aeli Donati commentum Terenti*, 3.1, *Eugraphi commentum*, ed. Paul Wessner [1908; repr. Stuttgart, 1963], p. 80), and *A Late Eighth-Century Latin–Anglo-Saxon Glossary*, ed. John Henry Hessels (Cambridge, Eng., 1906), p. 25 (in a series of 71 definitions of rhetorical figures taken from Cassiodorus's commentary on the Psalms).

[42] At least not in the 57 million words of texts currently scannable in the *Thesaurus Linguae Graecae*'s CD-ROM database.

Technology, Ecology and Religion:
Thoughts on the Views of Lynn White

by Bert Hall
and
Ranald Mackenzie Macleod

Lynn White, Jr., (1907–1987) was probably the most widely-read and influential medieval historian in the post-World War II American scene. His classic interpretive work, *Medieval Technology and Social Change*[1] remains indispensable to the study of all pre-modern technology. His essays and articles, collected as *Machina ex Deo: Essays in the Dynamism of Western Culture*[2] and *Medieval Religion and Technology*,[3] are likewise frequently cited. For many ecological activists and students of ecology however, the name of Lynn White is forever associated with only one idea, the "Lynn White Thesis," which claims that Judaeo-Christianity is at the root of the modern ecological crisis that threatens to engulf us all. The debate surrounding this thesis has been heated and has taken on a life of its own, continuing up to the present.[4]

Lynn White was a controversial and important intellectual figure, one whose ideas have always figured prominently in debates about the shape of medieval history. It is a tribute to the force of his thought that a broad and critical reconsideration of his legacy has begun to be carried out in the years since his death.[5] One of us (Hall) has recently published an article that seeks to explicate and place in context White's work up to the publication of *Medieval Technology and Social Change*.[6] The purpose of this essay is to place White's views on religion and ecology in the context of his later (post-1962) work, and to suggest that the argument White provoked risks losing sight of some of White's most important insights.

---

[1] Oxford, 1962.

[2] Boston, 1968. Subsequently reissued under the title *Dynamo and Virgin Reconsidered.*

[3] Publications of the Center for Medieval and Renaissance Studies, UCLA (Berkeley, Los Angeles, 1978).

[4] There is a critical review of the debate from a position largely hostile to White's thesis in Elspeth Whitney, "Lynn White, Ecotheology and History," *Environmental Ethics* 15 (1993), 151–69.

[5] White has recently been the subject of a master's thesis at the University of Oklahoma. See Judith Machen, "Cultural Values and the Vitality of the West: The Mind of Lynn White, Jr" (master's thesis, University of Oklahoma, n.d.). We are grateful to Mrs. Machen for sharing a copy of her study. See also *The Medieval West meets the Rest of the World: Essays to Honor the Memory of Lynn White, Jr,* ed. Nancy van Deusen, Claremont Cultural Studies (Ottawa, 1995).

[6] Bert Hall, "Lynn White's *Medieval Technology and Social Change* After Thirty Years," in *Technological Change: Methods and Themes in the History of Technology*, ed. Robert Fox (Australia, 1996).

White's intellectual career spanned the half century from the 1930s to the 1980s, during which time he served on the faculties of Princeton, Stanford and the University of California; as well, he was president of Mills College for some fifteen years (1943–58). Like any other thinker, White developed and changed during this time, and his thought needs to be seen in terms of continuities and discontinuities, phases and periods. Also, White was first and foremost an essayist, someone who felt most comfortable in the smaller confines of the academic article or learned essay. *Medieval Technology and Social Change* is his only "monograph" after his dissertation, *Latin Monasticism in Norman Sicily*.[7] It is necessary in any analysis of his work to keep in mind how he used the article format as a platform for his ideas and interpretations. This choice of formats focused his rhetoric and raised the temperature of debate in ways that are familiar enough in popular media culture, but which remain suspect in the eyes of academics. Indeed, reading Lynn White not infrequently puts one on the receiving end of an unforgettable and provocative set of ideas. The issues raised by these rhetorical strategies make it necessary to see White in the broader context he himself sometimes failed to provide.

This context begins with the understanding that Lynn White was a medieval cultural historian. Cultural history serves as the fundamental chord in White's *oeuvre*. Historians of technology tend to emerge from backgrounds in economic history, the history of science, or from the engineering disciplines themselves. White was none of these; he was, above all else, a medievalist, trained as an institutional and cultural historian. White went to Harvard in 1929 to study under Charles Homer Haskins,[8] the academic baron who fostered medieval studies in America during the 1920s. At first he did not aspire to emulate the Haskins of *Studies in Medieval Science* (1924), but Haskins the historian of Norman institutions. White's dissertation, completed under George LaPiana after Haskins' stroke rendered him an invalid, signals White's life-long interest in religious institutions and his tendency to see religious feeling as a primary motive to action.[9] Culture, for White, is grounded in religion, and religion serves as the bearer of cultural values.

After leaving Harvard for a teaching post at Princeton, White began his reading in the history of technology, first with Alfred L. Kroeber's *Anthropology*,[10] then with the more historiographically central writings of Franz Maria Felhaus, Marc Bloch and Richard Lefebvres des Noëttes. White was thus an autodidact, both with respect to technology itself as well as any possible interpretive frames within which to view technology. The attraction cultural historians feel for cultural anthropology is not unusual, of course. At a later stage, such feelings would produce the so-called "New Cultural History" that looks to

---

[7] Cambridge, Mass., 1938.

[8] On Haskins, see Norman Cantor, *Inventing the Middle Ages* (New York, 1991), pp. 245–286.

[9] Machen, "Values," pp. 31–32.

[10] New York, 1923.

Clifford Geertz and the method of "thick description" for its inspiration.[11] What emerges from the mix of anthropology, history and technology as these were read in the 1930s, however, is somewhat different.

The dominant characteristic in White's early thought was a buoyant optimism about the possibilities inherent in technology's advances, an attitude fully characteristic of the day. For many intellectuals during that troubled decade, technology was part of the saving remnant, something that offered hope when other human institutions had seemingly failed.[12] White extended this optimism in several directions. Like the newly-minted *Annales* school,[13] White argued that technology does not recognize conventional boundaries of class, geography or chronology. True to his roots as a cultural historian, he saw technology as coterminous with the great religions of the Old World: Christianity, Islam, Hinduism, Buddhism and Taoism. For White, technology was the artistic creation of the sub-literate masses of humanity, a universal and a unifying force in human affairs. Engineering, he later asserted, was the "new humanism" of our age.[14]

White added to all this the claim that medieval Europe played a critically important role in the development of Western technology. This was not entirely novel (Lewis Mumford, among others, had done much to popularize medieval Europe's contributions to technology), but it was still a far more radical claim in the 1930s and 1940s than it might seem today. In advancing such a claim, he followed a programmatic trend common to much of medieval studies in the United States. For Haskins and his intellectual heirs, the Middle Ages was always seen as historical terrain of strategic importance to the modern world. Breaking with the historiography of the Enlightenment, Haskins and his school insisted that the Middle Ages make sense, that there is a certain "rationality" to the medieval manner of doing things. One non-technological example of this approach by another of Haskins' students is found in Joseph Strayer's classic work on feudalism. In Strayer's reconstruction, feudalism is not an erratic, idiosyncratic collection of customs, but primarily a political and institutional form whose greatest virtue lies in its pragmatism: it works.

Another critical element of the Haskins' program is to deny any discontinuity between medieval Europe and the present. Modern states, modern ideas, modern

---

[11] Clifford Geertz, *The Interpretation of Cultures* (New York, 1973). On Geertz and the historians, see Lynn Hunt, et al., *The New Cultural History* (Berkeley and Los Angeles, 1989).

[12] Even Lewis Mumford, often taken as the archetype of intellectual pessimism concerning technology, was guardedly optimistic in the first redaction of his famous *Technics and Civilization* (New York, 1934).

[13] When Henri Pirenne died in the fall of 1935, Marc Bloch and Lucien Febvre devoted the memorial issue of the *Annales* to "Les Techniques, l'histoire et la vie." It was here that Bloch's famous essay on the medieval water mill appeared ("Avènement et conquêtes du moulin à eau," pp. 538–563), as well as essays by Febvre on the state of the field and desirable approaches to the history of technology ("Réflexions sur l'histoire des techniques," pp. 531–535; "Techniques, sciences et marxisme," pp. 615–623; and "Pour l'histoire de sciences et des techniques," pp. 646–648).

[14] "Engineers and the Making of a New Humanism," *The Journal of Engineering Education* 57 (1967), 375–376. Reprinted in *Machina ex Deo*, pp. 143–150.

institutions, modern habits of thought, and modern assumptions are all descended literally and lineally from a medieval European ancestry. Both the Renaissance and the Enlightenment saw modernity as born in the revolt against medieval ways; the new historiography (of which White was a significant part)[15] insisted on continuity between medieval and modern times. The modern world emerges continuously and predictably (if not always smoothly) from the womb of the Middle Ages. The Middle Ages made sense, and the modern world cannot be understood without reference to them.

Armed with these precepts, White worked for more than twenty years to explicate medieval technology to his humanistically-trained colleagues. From the publication of his pioneering essay in *Speculum* in 1940[16] to the appearance of *Medieval Technology and Social Change* in 1962, much of White's considerable intellectual energy went into this effort. White sometimes tried to fit technology into frameworks made of medieval social, political or economic history, but his primary concern was always to see technology as a cultural phenomenon and as an expression of deeper cultural values. *Medieval Technology and Social Change*, with its famous chapters dealing with the stirrup and feudalism and agrarian change and medieval demography, represents White's most strenuous attempt to integrate technology into medieval society, politics and economics; it was also his last serious attempt at this manner of argument. No sooner had he completed the manuscript than he returned to his root impulse, to explain medieval technology as a cultural and religious phenomenon.[17] Thereafter, right to the end of his scholarly career, he would devote all his efforts to what he saw as the more significant task before him.

The 1960s were, in general, a period of intense activity for White, but nothing ever matched the impact of his 1967 essay, published in *Science* under the title "The Historical Roots of Our Ecologic Crisis." Here White faced squarely the implications contained in virtually all his earlier work: if medieval technology is indeed of seminal importance for the modern world, and if Latin Christianity could somehow be credited with fostering the medieval West's technological dynamism, then religion must also share some of the responsibility for our culture's shameful treatment of the natural world. White took note of the environmental plight of contemporary industrialized societies and then drew up an indictment that surprised and outraged many. Even though it seems at first

---

[15] Cantor, *Middle Ages*, p. 285 "All these writers [chiefly White and Robert S. Lopez] were simply applying in extended areas of history of science, economy, and technology the primacy of rationality that Haskins and most eloquently Strayer perceived in medieval government, administration and law."

[16] "Technology and Invention in the Middle Ages," *Speculum* 15 (1940), 141–159. Reprinted in *The Making of Modern Europe*, ed. Herman Ausubel (New York, 1951), pp. 47–58. Also reprinted in *The Pirenne Thesis*, ed. A. F. Havighhurst (Boston, 1958), pp. 79–83, and in White, *Medieval Religion and Technology*, pp. 1–22.

[17] "In 1959, when I finished the manuscript of a book on medieval technology, I was painfully aware of its greatest defect: it . . . fails to explain the phenomenon observed." From the introduction to White's "Cultural Climates and Technological Advance in the Middle Ages," *Viator* 2 (1971), 171–201. Reprinted in White, *Medieval Religion and Technology*, pp. 217–254.

glance, he contended, that only modern scientific technology is running "out of control" in respect to its ecological effects, on deeper reflection, he concluded, the problem is much older, rooted in the Middle Ages and its dominant faith. "Christianity," he asserted, "bears a huge burden of guilt."[18]

White did not overlook the more commonly cited culprits in the environmental drama, science and industry. But he argued first that technology has a history of development that is independent from that of pure science. Only in the nineteenth century, he noted, did the marriage of science and technology change the essence of the relationship between humanity and the natural world. Science and technology were, White argued, both Occidental in origin and both nurtured in the Latin West. The nineteenth-century union of these cultural cousins gave bold new energy to a much older set of cultural trends, trends rooted in the twin Christian convictions of domination and exploitation. That is, the late twentieth-century environmental crisis could be traced back to a Christian concept of the relationship between humankind and nature, a concept that was born not in Apostolic times, but only in medieval Europe. White put it this way: "Christianity, in absolute contrast to ancient paganism and Asia's religions, not only established a dualism of man and nature but also insisted that it is God's will that man exploit nature for his proper ends."[19]

White's essay grew out of an invitation to address the American Association for the Advancement of Science on what was becoming the hottest topic of the 1960s: ecology. White's preparatory research did not yield any significant work on the history of ecology, and he was thrown back on his own resources.[20] He intended the A.A.A.S. talk as a challenge to some deeply-held beliefs about the character of modern technology, the ecological crisis it produces, and the entire body of assumption about the relationship between human beings and nature. Historiographically, the essay challenged the "modern" character of technology (i.e., the tendency to see technology as a child of science) in ways that are utterly consistent with the totality of White's other work.[21] There is, to be sure, a

---

[18] Lynn White, "The Historical Roots of Our Ecologic Crisis," *Science* 155 (1967), 1203–1207; quote, p. 1206.

[19] White, "Roots," p. 1205. It should be noted that White uses the word "religion" in two senses. First, in its commonly understood sense of a social institution, but secondly, and more importantly, as a set of fundamental ideas that colour one's philosophy. This incidentally also explains why it is so difficult to test White's thesis in any empirically convincing fashion. This has not stopped individuals from trying, though. For an interesting attempt, see Douglas Lee Eckberg and T. Jean Blocker, "Varieties of Religious Involvement and Environmental Concerns: Testing the Lynn White Thesis," *Journal for the Scientific Study of Religion* 28 (1989), 509–517.

[20] Clarence Glacken, *Traces on the Rhodian Shore: Nature and Culture in Western Thought from Ancient Times to the End of the Eighteenth Century* (Berkeley and Los Angeles, 1967) appeared later in the same year as White's article.

[21] White can be said to have anticipated his own later work in his essay "Naturalistic Science and Naturalistic Art in the Middle Ages," *American Historical Review* 52 (1947), 421–435. In the final paragraph of that essay he concludes that "modern science was one result of a deep-seated mutation in the general attitude towards nature, of the change from a symbolic-subjective to a naturalistic-objective view of the physical environment."

dramatic difference in tone between "Roots" and much of White's earlier work: where White was once mainly optimistic about the possibilities for a technological future, he was now plainly much more worried. We might recall at this point just how deeply disturbing the 1960s were, especially in California's perpetually sunny and upbeat climate. The American high-tech cheerfulness that sent men to the moon and made "war on poverty" was quickly crumbling in the face of assassinations, riots, the Vietnam War, and an apocalyptic sense of impending ecological disaster. Inevitably, the 1960s became trivialized as they were made into pop-culture nostalgia; to those of us who lived them, they were a turning point of almost unimaginable significance.

White's shift of tone brought him into confrontation with a solid block of contrary-minded opinion, and the theological aspect of the debate quickly came to occupy center stage. Few objected when White found Latin Christian roots for the West's triumphs in technology, but when White's intellectual honesty demanded that our religious attitudes bear some measure of responsibility for the ecological consequences of that technology, there were protests aplenty. White faced a growing chorus of hostile responses from academics, and, for the first time in his life, hate mail.[22]

White's essay, of course, was not the first attempt to weave religion into the fabric of intellectual debate over technology and culture. Max Weber's 1904 *The Protestant Ethic and the Spirit of Capitalism* had attempted to link the Protestant faith with modern styles of economic production,[23] while only ten years before "Roots" appeared, John Nef had argued in his *Cultural Foundations of Industrial Civilization* that methods and insights derived from theology had helped to advance scientific and technological knowledge.[24] In fact, it appears that White was not even the first to formulate the thesis that is commonly associated with his name. The Zen Buddhist scholar, D. T. Suzuki, outlined essentially the same argument in 1953.[25] Perhaps because it was uttered at the right time and place, the "Lynn White Thesis" has gone on to become a virtual cliché. It has become, in the words of one recent critic, both "overwhelmingly familiar" and "a fruitful touchstone for debate."[26] Yet it remains for many others a suspicious and unsatisfactory argument.

Debate about the "Lynn White Thesis" has often taken place in contexts that give little attention to the lifetime of work that bore fruit in his observation. This

---

[22] "I was denounced, not only in print but also on scraps of brown paper thrust anonymously into envelopes, as a junior Anti-Christ, probably in the Kremlin's pay, bent on destroying the true faith." Lynn White, Jr., "Continuing the Conversation," in *Western Man and Environmental Ethics: Attitudes Toward Nature and Technology*, ed. Ian G. Barbour (Reading, Mass., 1973), pp. 55–64; quote, p. 60.

[23] Max Weber, *The Protestant Ethic and the Spirit of Capitalism*, trans. Talcott Parsons (London, 1930).

[24] John Nef, *Cultural Foundations of Industrial Civilization* (Cambridge, Eng., 1958), esp. pp. 36–64.

[25] "The Role of Nature in Zen Buddhism," *Eranos-Jahrbuch 1953* 22 (1954), 291–321. Cited in Glacken, *Traces*, p. 494.

[26] Whitney, "Lynn White," p. 151.

has often led to criticisms that either miss the mark completely, or seem designed to twist the argument (not surprisingly) toward the critic's field of expertise. It seems altogether too self-evident that modern technological society has acted with neither restraint nor delicacy in its manipulation of the material world. Broad agreement on this point means that the remaining questions tend to become normative. What should the relationship between humanity and nature be? Who, or what, should accept blame for the present state of affairs? This is the emotional "fuel" for the debate, the search for someone or something to "blame" for the manifestly exploitative attitudes that modern culture manifests toward nature, or, alternatively, the effort to exempt from possible blame belief systems or institutions held dear.[27]

Arguably, such an approach grasps the wrong end of the stick. After making allowances for White's 1960s audience and its preconceptions, his rhetoric should not be seen as an attempt to cast blame, and certainly not as a search for a convenient scapegoat. Rather, he is engaged in a very difficult task for the cultural historian: pointing to an absence. The Western religious tradition has little built-in respect for nature *as such*, seeing in nature either the home of the enemy cult[28] or a mere material resource to be harnessed and put to work towards more spiritual ends. How does one make an audience aware of an attitude so deeply held that it almost defies verbalization, much less analysis? "Roots" reminds the reader of what makes European Christianity unique, and it suggests alternatives that may deserve a fuller hearing than they have heretofore received.

White's emphasis on religion as a driving force in history and on the importance of medieval Europe makes several critics argue that he takes too limited a view.[29] The issue seems to be the origins of modernity itself; if becoming modern is seen in its traditional guise as a revolt against traditional ("medieval") patterns, then so, too, ecological degradation (that most modern of modernity's problems) must be rooted in the secular revolt against religious tradition which many still believe characterizes all post-medieval societies.[30] It is difficult at times to know what to make of such a claim. It contradicts the efforts of several generations of medievalists—not the least of them Lynn White himself—to show how the medieval and the modern worlds are related. For many of us, to reject that claim is to impoverish our sense of history itself. There is, of course, no final rebuttal, no conclusive riposte, to this insistence that the Middle

---

[27] See, for example, Katherine Temple, "Doubts Concerning the Religious Origins of Technological Civilization," and Mark Swetlitz, "A Jewish Commentary on the Religious Origins of Technological Civilization," both in *Research in Philosophy and Technology*, vol. 6, ed. Paul T. Durbin (Greenwich, Conn., 1983), pp. 189–197 and 197–203 respectively.

[28] Much work could be done, it seems, on the antipathy between Semitic monotheism and the fervent nature cults of the Ancient Near East, an antagonism that was recapitulated in barbarian Europe as Christian monks carried out missions to "pagan" peoples.

[29] See, for example, Lewis Moncrief, "The Cultural Basis for Our Environmental Crisis," *Science* 170 (1970), 508–512.

[30] See Carl Mitcham, "The Religious and Political Origins of Modern Technology," in *Philosophy and Technology*, ed. Paul T. Durbin and Friedrich Rapp (Dordrecht, 1983), pp. 267–273.

Ages remain irrelevant to the modern condition. One can only shrug and conclude that the argument has passed beyond the realm of the rational. All cultures select what they wish to emphasize from the rich possibilities inherent in the past as they construct their myths of themselves. All the historian can do is to make the case for a period's significance. One is, however, reminded of the apocryphal man who, speaking Esperanto, found it impossible to believe that the world did not begin in the twentieth century.

It might help to note in this respect that, although White was convinced that religion and the Middle Ages are both of great importance in the modern world, his thesis does nothing to pre-empt any "modernization" argument. It merely points to an additional—medieval—root of the problem. At no time does White declare the issue closed to further investigation. In the preface to his *Medieval Religion and Technology*, White notes that the technological thrust of the medieval West does not yield easily to explanation, and that religion was only the most powerful of the several forces that shaped the Middle Ages. In his reply to critics of "Roots," he makes much the same point: "It is this sense of pluralism, and the various strata of historical 'causation,' that lead me to prefer the metaphor of roots."[31] On the whole, his is a much more accommodating, almost more conservative, stance than those of his critics.

A second strain of criticism accuses White of a form of historical myopia, of failing to recognize that environmental change has been occurring at least since antiquity. There is some merit to this claim, but it reflects a certain fashionable weakness in academic culture for what we might call "universals," and a concurrent tendency to make all claims relative. Assaults on the natural environment become, in this view, merely part of "human nature" and thus presumably unavoidable. "Everybody's doing it," says this line of criticism. "Why should we be any different?"

Such an argument blurs important distinctions, as does any similar appeal to a universal "human nature." Killing other human beings might well be endemic in all (or virtually all) human cultures, but there are surely important differences between cultures on just what constitutes "murder" and how tolerant we should be towards killing members of our own species. These cultural differences help shape the daily lives of citizens of such different cities as, let us say, Toronto, Detroit, or Sarajevo. Erotic desire, too, is doubtless fairly evenly distributed among people of all races and cultures, but what constitutes "rape" or other prohibited expressions of sexual activity depends largely on local laws, customs, and traditions. Universalizing an issue often inhibits our ability to place it in context. White began his essay with the note that "all forms of life modify their contexts,"[32] but he did not seek to establish a taxonomy or a genealogy of environmental assaults. He does not, in fact, distinguish between degradation due to simple over-population and that stemming from over-consumption, nor does he seek to characterize some forms of degradation as Western and other forms as non-Western.

---

[31] White, "Continuing the Conversation," p. 57.
[32] White, "Roots," p. 1203.

But none of this was his central task in what was, after all, a 1967 essay based on an address, not a monograph.[33] It was never White's aim to provide a comprehensive overview of environmental history. Instead, he asked us to regard *ourselves*, and to contemplate why *we* regard nature with indifference. His use of non-European examples was confined to the task of making us aware of the lacuna in our customary spirituality where nature is concerned. The issue is complicated, to be sure, by the fact that modern non-Western forms of environmental damage are often connected with the export to non-Western settings of Western science and technology. (Japan comes immediately to mind.) Here, one must admit, White simply assumed that environmental problems result from the absorption of occidental attitudes (i.e. Christian attitudes dressed up as "modern, scientific" views) along with Western science and technology. This is a complex subject, and worthy of a considerable effort to elucidate,[34] but it should not be made a convenient club with which to cudgel White.

A third line of criticism focuses on White's interpretation of Christian moral values vis-à-vis nature, and specifically on what his critics see as his tendency to ignore the biblical mandate enjoining "stewardship"[35] as the ideal model for relations between humankind and nature. This is a serious charge, and one that has implications for White's other work as well. It turns on a subordinated methodological criticism that few of White's critics have made very explicit: a rejection of White's reliance on what he termed "sub-verbal" attitudes as key elements in determining a culture's real approach to nature. These two criticisms intertwine, as theologically sophisticated intellectual historians seek to rebut White's reading of medieval history while substituting their own theologically "sanitized" reading of earlier attitudes. White's method of using resources was always controversial; iconography, manuscript illuminations, homiletic tropes and images, artifacts and the like—all were mined to explicate in very broad ways the generic attitudes of medieval cultures toward technology, work and nature. Carl Mitcham responds to White's methodological novelty with the

---

[33] Historical work done after 1967 fills in some of the record here. Yi-Fu Tuan, "Our Treatment of the Environment in Ideal and Actuality," *American Scientist* 58 (1970), 244–277, covers similar ground to White's with an emphasis on Asia. Two other useful works are Clive Ponting, *A Green History of the World* (London, 1991) and Andrew Goudie, *The Human Impact* (Oxford, 1981). Ponting in particular provides strong arguments that human activities, especially agriculture, have wrought destructive changes on the natural landscape for much, if not all, of the span of recorded history.

[34] Although the full extent of environmental damage in the formerly Communist world was not widely known in 1967, it is interesting to observe that more recent knowledge of those regimes serves to confirm White's point, regardless of the political or economic system under consideration. Communism, despite its claims to atheism, environmentalism, and worker's rights, served both nature and workers rather badly. The environmental record of former Soviet satellites is, quite simply, ghastly. In her recent critique, Whitney appeals for increased study of economic systems, but simple observation would seem to indicate that environmental distress is a symptom of more than just economics.

[35] White, "Continuing the Conversation," p. 60: "The most common charge [by critics] was that I had ignorantly misunderstood the nature of 'man's dominion' and that it is not arbitrary rule but rather a stewardship of our fellow creatures for which mankind is responsible to God."

strangely old-fashioned claim that White seems unable "to appreciate the difference between knowledge and opinion, the thought of the few versus that of the many."[36]

Many scholars find White's "anthropological" approach distasteful, and challenged White to provide evidence they would regard as more compelling, i.e. textual sources. One might respond to such a challenge by arguing that, as the subject of technology itself is only weakly represented in medieval texts produced by scribal cultures remote from the gritty business of daily life, so too the attitudes of the unlettered masses are even more poorly expressed. What sort of view of twentieth-century America might a future historian gather if his sources were limited to back issues of *Christian Century* and *Partisan Review*? Mitcham in fact represents a strain of academic elitism masked as methodological conservatism. All cultures are much more than the mere sum of their textual products, and the attitudes of those who express themselves in deeds rather than words *are* important, just as important as the attitudes of those who, like ourselves, employ words. Surely, as White argued throughout his career, we must reach deeper than the merely literate if we wish to understand some valuable aspects of medieval attitudes.

The juncture between "popular" and "elite" cultures is exactly where the problem of attitudes needs the most clarification. White's argument was widely misunderstood on just this point. He did not contend that Christianity as a whole somehow systematically *advocated* the exploitation of nature, nor that learned commentaries specifically endorsed environmental rape. He was precisely concerned that Western, Latin-speaking Christian Europe had *subverted* Scriptural precedent to accommodate exploitative attitudes. Scripture, White maintained, had indeed created a division between humankind and nature, a contention so obvious it can scarcely be debated. White pointed to Genesis 1:28, where God commands man to

> be fruitful, and multiply, and replenish the earth, and subdue it: and
> have dominion over the fish of the sea, and over the fowl of the air, and
> over every living thing that moveth upon the earth.

This and similar passages convinced White that the Judaeo-Christian tradition contained within its primal myth the seeds of exploitation; humans were created in God's image, and from God, humans have a mandate to dominate the world of nature. At the same time, White recognized that this conclusion would be, as he put it, "unpalatable to many Christians," and he sought to modify the Scriptural simplicity of this position by arguing just as strongly that it was only in the Latin West that this seed sprouted and grew. Indeed, one of the recurrent themes in White's thought concerns the "failure" of Orthodoxy, Judaism or Islam to follow the Western path towards high technology (and thus towards environmental depredation). It is this distinction between Western and non-Western forms of

---

[36] Mitcham, "Origins," p. 272.

Christianity that sets White's views apart from those of Ernst Benz,[37] who did indeed make the sorts of sweeping generalizations about the entire Christian tradition that so many critics have attributed to White.

By focusing too much on the biblical theme of stewardship, the response to White on these points becomes mired in the Is-Ought fallacy. Robin Attfield's treatment may be taken as representative. White, he begins, is "wrong to hold that Christians have usually held that people may treat nature as they please." He then proceeds to outline the stewardship argument:

> The biblical position, which makes people responsible to God for the uses to which the natural environment is put, has never been entirely lost to view, and may properly be appealed to by the very people who rightly criticize the exploitative attitudes which prevail in many places throughout the contemporary world.[38]

As an alternative to White's "domination and exploitation" exegesis, he offers up a different passage from Genesis: "And the Lord God took the man, and put him into the garden of Eden to dress it and to keep it."[39] Attfield reads this to mean that "man was thus to preserve the garden's beauty and protect it from harm, as well as derive his food from it."[40] The problems inherent in this reading may not be obvious simply because the attitude Attfield expresses is so ubiquitous. Consider the word "dress." It carries an added meaning, "to fashion" or "to prepare." Likewise the injunction to keep the Garden, though it might suggest preservation, is expanded by Attfield, as it would be by most commentators, to suggest that man is to employ the Garden as a food source, and that its "beauty" (surely a product of the human mind as much as of nature) is a salient characteristic. In short, under Attfield's interpretation, nature exists to serve humans and human interests.

The only accommodating gesture Attfield makes in the direction of White's thesis is equally revealing. He concedes that "there is some justification for highlighting what needs to be rejected if our attitudes are to be wholesome ones."[41] But a search for "wholesome" attitudes (whatever they might be) does not move beyond the "prudential" view that nature's principal purpose is to serve human needs. In the triadic relationship between God, human beings, and nature, nature itself has no spiritual value, merely instrumental utility. White's argument, both in its historical and its theological aspects, contends that Western Christians have indeed so thoroughly desacralized the natural that we are no longer able to regard nature with much sense of reverence, and we cannot begin to conceive that

---

[37] See Ernst Benz, "Fondamenti cristiani della tecnica occidentale," in *Tecnica e casistica: tecnica, escatologia e casiastica*, ed. Enrico Castelli, Convegno internazionale di studi umanistici 9 (Rome, 1964), pp. 241–63.

[38] Robin Attfield, "Christian Attitudes Toward Nature," *Journal for the History of Ideas* 44 (1983), 369–86; quote, p. 386.

[39] Ibid., p. 374.

[40] Ibid., p. 374.

[41] Ibid., p, 369.

any natural creatures other than ourselves might have "rights." White rejected the image of earth as seen from space, an image made popular in the 1960s, in which our planet itself is seen as a spaceship for the entire human race. To him, the spaceship metaphor is the ultimate extension of a prudential view of nature, and the "terrifying" extinction of any remnant of nature's right to exist for the sake of nature, not for human purposes.[42]

In the end, the issue is not to be joined on the plane of fine-tuned Scriptural exegesis. There can be no doubt that the stewardship argument represents the best that the Judaeo-Christian tradition can offer with respect to protection of nature, and it is perhaps pointless to expect even a professedly post-Christian culture to move beyond the limits of prudentialism. But if one wants to be polemical at this point, it would be easy to point out how feeble the stewardship argument is in practice. One might sarcastically ask where stewardship ranks in the array of virtues, since neither Scripture nor theology pays much attention to it. The punitive and wrathful God of the Old Testament shows a singular lack of interest in the environment; despite a plethora of moral examples where many other sins and sinners are singled out for punishment, in no instance does Yahweh choose to punish anyone for crimes against the environment. We are not surprised when American fundamentalists organize letter-writing campaigns denouncing Apple Computer Corporation for extending family benefits to gay or lesbian employees. We *would* be surprised, however, if these same fundamentalists were to chain themselves to trees (or even write letters to legislators) demanding an end to the logging of old-growth forests. James Watt, Ronald Reagan's first Interior Secretary, responded to a Congressional Committee's questions about the value of preserving natural sites for future generations with the alarming comment, "I do not know how many future generations we can count on before the Lord returns."[43] Moderate Christians may find such a statement shocking, even scandalous, but in its guileless and unabashed zealotry Watt's words reveal an attitude that lurks within many believers, albeit in more shaded and nuanced forms. What can the ultimate value of nature be in a metaphysical system which insists humankind's ultimate locus lies elsewhere? Recognition of these facts explains why attempts to deal with White's argument through finely-tuned Scriptural exegesis generally fall short of the mark; exegesis does not "know" practice—or popular culture, for that matter.

A first step towards moving past sterile debate surely must be to recognize that Christianity has often been a pragmatic exercise in accommodating belief to experience. The uses to which Scripture has been put have varied with time and place, while adherence to Scriptural injunctions has been, and remains, subject to a great deal of interpretive latitude. In this sense then, the claims of White's theological critics may be quite correct in ways they did not intend; in the latter half of the twentieth century we may be on the verge of a radical revision of the theology of nature. Justification for a benevolent approach to nature may be, at last, not merely found in Holy Writ, but actually taught and *believed* to be the

---

[42] White, "Continuing the Conversation," pp. 63–64.
[43] Grace Haskell, *Prophecy and Politics* (Westport, Conn., 1986), p. 8.

most appropriate way to mend the relationship between God, humankind, and nature. Paradoxically, though, White's claim loses none of its own vigor by recalling again one of its central points, that it is Western culture, not Scripture itself, that sees no value to nature other than our own utility. The historian is accustomed to situations where texts used for centuries to justify certain attitudes can receive profoundly altered readings in quite short periods of time. White sees in this a ray of hope. He notes:

> Scattered through the Bible, but especially the Old Testament, there are passages that can be read as sustaining the notion of a spiritual democracy . . . of all creatures. The point is that historically they seem seldom or never to have been so interpreted. This should not inhibit anyone from taking a fresh look at them.[44]

"Taking a fresh look" is in fact just what has occurred since White's essay was first published. Recent years have seen an upsurge of interest in ecotheological questions, that is, in questions of "right livelihood," of proper human conduct towards the natural world. Outside the sometimes cramped world of Christian theology, philosophical activity has also developed its own momentum, so that there now exists a broad spectrum of opinion on matters of culture, technology, and nature. Two groups currently representing extreme positions with regard to the philosophy of technology and nature are the "deep ecologists" and the "technological optimists." The "deeps" maintain that all life forms are of equal intrinsic value to human beings, and reject attempts to protect only those elements of nature that are seen as being of service to humans. The "optimists," on the other hand, recognize no such limits to the technological endeavor; human ingenuity, they assume, will discover new "resources" as materials currently in use become increasingly scarce.[45]

Lynn White was deeply gratified by the intellectual movement these debates represent. He was, when all is said and done, no technological determinist. Recall his essay's final prescription: "More science and more technology are not going to get us out of the present ecologic crisis until we find a new religion, or rethink our old one."[46] He always emphasized the necessity of spiritual approaches. This is why he promoted humility in the face of nature as the most useful guide for human behavior. Some critics, such as Whitney, feel this insistence merely encourages personal inactivity, though; it "requires no immediate action" on the

---

[44] White, "Continuing the Conversation," p. 61. White cites slavery by way of comparison. Once nearly universally believed moral and supported by Scripture, slavery underwent a complete transvaluation in little more than a century between c. 1730 and 1865.

[45] The term "deep ecology" was coined by the Norwegian ecophilosopher Arne Naess; his *Ecology, Community, and Lifestyle: Outline of an Ecosophy*, trans. David Rothenberg (Cambridge, Eng., 1989) is a good introduction to the topic. A less rigorous treatment can be found in Bill Devall and George Sessions, *Deep Ecology: Living as if Nature Mattered* (Salt Lake City, 1985). The most powerful (some would say the most outrageous) statement of the "optimist" view is given by Julian Simon in his *The Ultimate Resource* (Princeton, 1981).

[46] White, "Roots," p. 1205.

individual's part.[47] No one who knew Lynn White would ever have agreed; for White, spirituality was the source and the guide to all action. Whether one agrees with White's prescription or not, one must recognize that his essay has had lasting value. Its key insight, the realization that the environmental crisis is a cultural phenomenon, not merely a technical one, opened the door to an entirely new realm of philosophical investigation. A generation after the publication of "The Historical Roots of Our Ecologic Crisis," it is evident that both "deep ecologists" and "technological optimists" owe an intellectual debt to Lynn White and his provocative ways.

UNIVERSITY OF TORONTO

---

[47] Whitney, "Lynn White," p. 168.